Jenny Sharp
Stewart Towne
Series Editor:
Jayne de Courcy

AS Maths

Contents

Published by HarperCollins*Publishers* Ltd
77-85 Fulham Palace Road
London W6 8JB

First published 2002

ISBN 0 00 712427 9

Jenny Sharp and Stewart Townend assert the moral right to be identified as the authors of this work.

British Library Cataloguing in Publication Data
A catalogue record for this book is available from the British Library.

Edited by Joan Miller
Production by Kathryn Botterill
Design by Gecko Ltd
Illustrations by Gecko Ltd
Screen dumps provided by the authors
Cover design by Susi Martin-Taylor
Printed and bound by Printing Express Ltd, Hong Kong

Every effort has been made to contact the holders of copyright material, but if any have been inadvertently overlooked, the Publishers will be pleased to make the necessary arrangements at the first opportunity.

Get the most out of your Instant Revision pocket book

1 **Maximize your revision time.** You can carry this book around with you anywhere. This means you can spend any spare moments revising.

2 **Learn and remember what you need to know.** The book contains all the really important facts you need to know for your exam. All the information is set out clearly and concisely, making it easy for you to revise.

3 **Find out what you don't know.** The *Check yourself* questions help you to discover quickly and easily the topics you're good at and those you're not so good at.

What's in this book

1 The content you need for *your* AS exam

- To achieve an AS in Mathematics you require three modules. This book covers four of the AS Mathematics modules: Pure 1, Pure 2, Mechanics 1 and Statistics 1. You should be doing at least two of these modules in your AS Mathematics.
- The 'mathematics' sections, i.e. Pure 1 and Pure 2, are put together in this book to provide a coherent structure. If you are not studying Pure 2, there will be parts which you have not studied yet.

2 Worked examples

- Mathematics is a subject that is often best explained through an example. All the topics in this book contain examples with full worked solutions, so you can see the mathematics you need to use to solve the problems. Remember that mathematics is not a set of discrete sections: to solve many problems you will need to rely on knowledge from previous topics. Remember too that mathematics is best learnt by actually doing it. Simply reading through worked examples may not be very beneficial. You can use the worked examples as revision questions and see how your solution matches the given one.

3 Formulae and equations

- There are a number of formulae and equations that you will be required to remember for your examination. This book contains all the important facts that you need to learn.

4 *Check yourself* questions – find out how much you know and improve your grade

- The *Check yourself* questions occur at the end of each short topic.
- The questions are fairly quick to answer. They are not actual exam questions, but the authors have written them in such a way that they will highlight any vital gaps in your knowledge and understanding.
- The answers are given at the back of the book. When you have answered the questions, check your answers with those given. The authors give help with arriving at the correct answer, so if your answer is incorrect, you will know where you went wrong.

Revise actively!

- **Concentrated, active revision** is much more successful than spending long periods reading through notes with half your mind on something else.
- The sections in this book are quite short. For each of your revision sessions, choose a couple of topics and concentrate on them for **20–30 minutes**. You should work through the examples yourself. Then do the *Check yourself* questions. If you get a number of questions wrong, you will need to return to the topics at a later date. You may find that you need to revise some topics a few times but, by coming back to them several times, your understanding will improve and you will become more confident about using them in the exam.
- Use this book to revise either on your own or with a friend!

Solving equations

An **equation** is a mathematical statement that two expressions are equal. You can manipulate an equation to solve it but you must always apply the same operation to both sides of the equals sign, to keep the balance.

Q Solve $12 - 3x = x + 4$

Add $3x$ to both sides: $12 - 3x + 3x = x + 4 + 3x \Rightarrow 12 = 4x + 4$

Subtract 4 from both sides: $12 - 4 = 4x + 4 - 4 \Rightarrow 8 = 4x$

Divide each side by 4: $\frac{8}{4} = \frac{4x}{4} \Rightarrow 2 = x$

Sometimes you will need to solve an equation that includes brackets that must be **expanded** (multiplied out).

Q Solve $3(x - 4) = 5(x + 4)$

First expand the brackets on both sides: $3x - 12 = 5x + 20$

Add 12 to both sides: $3x = 5x + 32$

Subtract $5x$ from both sides: $-2x = 32$

Divide each side by -2: $x = -16$

Sometimes the equation will contain fractions. You need to multiply each side by the lowest common denominator to eliminate the fractions.

Q Solve $\frac{3x + 2}{2} = \frac{x - 1}{3}$

Multiply each side by 2×3 to remove the fractions:

$$\frac{6(3x + 2)}{2} = \frac{6(x - 1)}{3} \Rightarrow 3(3x + 2) = 2(x - 1)$$

Now expand the brackets on both sides: $9x + 6 = 2x - 2$

Subtract 6 from both sides: $9x = 2x - 8$

Subtract $2x$ from both sides: $7x = -8$

Divide each side by 7: $x = -\frac{8}{7}$

Q Solve $\frac{4 + x}{x} = \frac{3}{5}$

Multiply each side by $5x$ to remove the fractions:

$$\frac{5x(4 + x)}{x} = \frac{5x \times 3}{5} \Rightarrow 5(4 + x) = 3x$$

Now expand the brackets: $20 + 5x = 3x$

Subtract 20 from both sides: $5x = 3x - 20$

Subtract $3x$ from both sides: $2x = -20$

Divide each side by 2: $x = -10$

Simplifying expressions

When the solution to a problem gives an algebraic expression, always express it in its **simplest form**. This is good mathematical practice!

Q Simplify $(8x + 7y) - (2y + 5x)$

You need to take out the brackets and just leave an x-term and a y-term.

Remove the brackets: $8x + 7y - 2y - 5x$

Collect like terms: $3x + 5y$

Q Simplify $7x(2x - 3) + 4(2x - 5)$

Multiply out brackets: $14x^2 - 21x + 8x - 20$

Collect like terms: $14x^2 - 13x - 20$

Q Simplify $5x(5x + 2) - 4(5x - 5)$

Multiply out brackets: $25x^2 + 10x - 20x + 20$

Collect like terms: $25x^2 - 10x + 20$

Take out a common factor: $5(5x^2 - 2x + 4)$

Sometimes an expression is easier to work with if it has been **factorised** (i.e. put into brackets).

Q Simplify $24q^2 - 30q - 42qp^2$

q is a common factor to all terms: $q(24q - 30 - 42p^2)$

6 is a common factor to all terms: $6q(4q - 5 - 7p^2)$

Changing the subject of a formula

A **formula** is a **mathematical statement** for working out the value of a quantity. For example, a formula used in the study of elastic strings in mechanics is $T = kx + e$. You can **change the subject of the formula** by rearranging to express x in terms of T. You can treat a formula as an equation and solve it for the required variable.

Q Make x the subject of $T = kx + e$

Treat the formula as an equation and solve for x.

Subtract e from both sides: $T - e = kx$

Divide both sides by k: $\frac{T - e}{k} = x$

Swap sides so x is on the left: $x = \frac{T - e}{k}$

Q Make a the subject of the formula $v = u + at$

Subtract u from both sides: $v - u = at$

Divide each side by t: $\frac{v - u}{t} = a$

Swap sides so a is on the left: $a = \frac{v - u}{t}$

Q Make x the subject of the formula $y = \frac{x + 4}{2x + 1}$ and find the value of x when $y = 1$.

Multiply both sides by $(2x + 1)$ to remove the fraction:

$$y(2x + 1) = x + 4$$

Expand the bracket: $2xy + y = x + 4$

Subtract 4 from both sides: $2xy + y - 4 = x$

Collect all terms containing x on the right-hand side:

$$y - 4 = x - 2xy$$

Take x out as a common factor: $y - 4 = x(1 - 2y)$

Divide both sides by $(1 - 2y)$: $\frac{y - 4}{1 - 2y} = x$

Swap sides so x is on the left: $x = \frac{y - 4}{1 - 2y}$

So when $y = 1$: $x = \frac{1 - 4}{1 - 2 \times 1} = \frac{-3}{-1}$

$$= 3$$

Linear and quadratic inequalities

A linear inequality has the form $3x + 2 < x - 1$ or $3x + 2 \geqslant x - 1$.

A quadratic inequality has the form $2x^2 - 5x - 3 \leqslant 0$.

You can simplify inequalities, following these three rules.

- Adding or subtracting the same positive or negative quantity to both sides leaves the inequality unchanged.
- Multiplying or dividing both sides by the same positive quantity leaves the inequality unchanged.
- Multiplying or dividing both sides by the same negative quantity changes the sense of the inequality (from $<$ or $\leqslant$ to $>$ or $\geqslant$ and vice versa).

Q Solve the inequality $3x + 2 < x - 1$

Subtract 2 from both sides: $3x < x - 3$

Subtract x from both sides: $2x < -3$

Multiply both sides by $\frac{1}{2}$: $\frac{1}{2} \times 2x < \frac{1}{2} \times -3 \Rightarrow x < -\frac{3}{2}$

Q Solve the inequality $|2x + 3| \leqslant 7$

This problem needs to be tackled in two stages depending on whether $2x + 3$ is positive or negative.

- If $2x + 3$ is positive, $|2x + 3| = 2x + 3$ and so we have $2x + 3 \leqslant 7$.
  First $2x + 3 > 0 \Rightarrow x > -\frac{3}{2}$ and $2x + 3 \leqslant 7 \Rightarrow x \leqslant 2$.
  So the solution must satisfy both these i.e. $-\frac{3}{2} < x \leqslant 2$.
- If $2x + 3$ is negative, $|2x + 3| = -(2x + 3)$ and so we have $-2x - 3 \leqslant 7$.
  Now $2x + 3 < 0 \Rightarrow x < -\frac{3}{2}$ and $-2x - 3 \leqslant 7 \Rightarrow x \geqslant -5$.
  So the solution must satisfy both of these, i.e. $-5 \leqslant x < -\frac{3}{2}$.

So the overall solution must satisfy both conditions, i.e.$-5 \leqslant x < 2$.

Q Solve the inequality $2x^2 - 5x - 3 \leqslant 0$

Factorising gives:
$(2x + 1)(x - 3) \leqslant 0$

The roots of the equation are $x = -\frac{1}{2}$ and $x = 3$.

From a sketch, the solution is $-\frac{1}{2} \leqslant x \leqslant 3$.

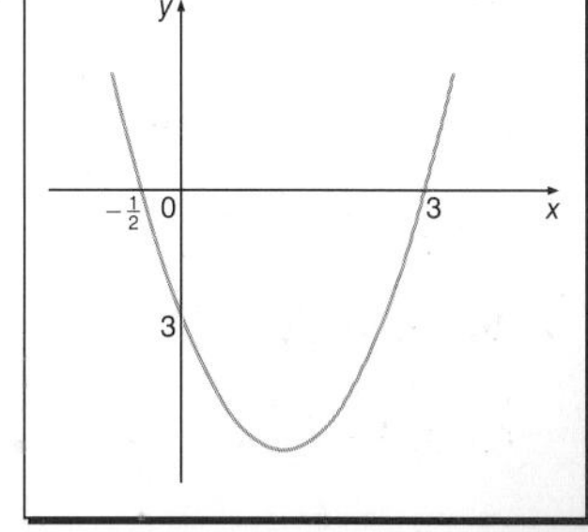

Laws of indices

You need to know the four laws of indices (or powers).

For any positive number a, and any numbers m and n:

1 $a^m \times a^n = a^{m+n}$

2 $\frac{a^m}{a^n} = a^{m-n}$

3 $(a^m)^n = a^{mn}$

4 $a^0 = 1$

Remember that $a^{\frac{1}{2}} = \sqrt{a}$, $a^{\frac{1}{3}} = \sqrt[3]{a}$ and $a^{-1} = \frac{1}{a}$, $a^{-2} = \frac{1}{a^2}$ and so on.

Q Simplify:

(a) $\frac{4^3 \times 4^5}{4^6}$ **(b)** $3x^3 \times 2x^4$

(c) $(4ab^2)^2$ **(d)** $\frac{3p^5q^2r^4}{9p^4q^3r}$

(a) Apply rule 1: $\frac{4^{3+5}}{4^6} = \frac{4^8}{4^6}$

and then rule 2: $4^{8-6} = 4^2 = 16$

(b) Apply rule 1 to the powers of x:
$3 \times 2 \times x^{3+4} = 6x^7$

(c) Expand the brackets: $4^2 \times a^2 \times (b^2)^2$
and then apply rule 3: $16a^2b^4$

(d) Apply rule 2 to the quotients involving p, q and r separately:
$\frac{3}{9}p^{5-4}q^{2-3}r^{4-1} = \frac{1}{3}pq^{-1}r^3 = \frac{pr^3}{3q}$

Q Simplify:

(a) $\sqrt{\frac{16p^8}{9q^2}}$ **(b)** $\sqrt{2\frac{7}{9}}$ **(c)** $(2\frac{7}{9})^{\frac{3}{2}}$

(a) Rewrite the square root as:

$$\frac{\sqrt{16}\sqrt{p^8}}{\sqrt{9}\sqrt{q^2}} = \frac{4(p^8)^{\frac{1}{2}}}{3q} = \frac{4p^4}{3q}$$

(b) $\sqrt{2\frac{7}{9}} = \sqrt{\frac{25}{9}} = \frac{5}{3}$

(c) $(2\frac{7}{9})^{\frac{3}{2}} = \frac{25^{\frac{3}{2}}}{9^{\frac{3}{2}}} = \frac{(25^{\frac{1}{2}})^3}{(9^{\frac{1}{2}})^3} = \frac{5^3}{3^3} = \frac{125}{27}$

Surds

A **surd** is the **square root** of a **positive number**, expressed in the form $\sqrt{x}$ without being evaluated numerically. For example $\sqrt{5}$ denotes the exact value of the square root of 5, as opposed to the approximate value given by a calculator. You can use the usual arithmetic operations to manipulate surds.

Q Determine the length of the diagonal of a rectangle 4 cm by 8 cm.

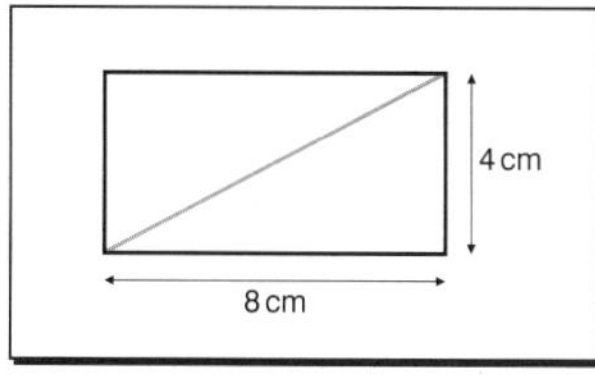

By Pythagoras' theorem, the length of the diagonal is:
$\sqrt{4^2 + 8^2} = \sqrt{80} = \sqrt{16 \times 5} = 4\sqrt{5}$.

This is the exact value of the length of the diagonal.

Q Simplify $8\sqrt{5} - \sqrt{125} + \sqrt{80}$

The aim is to reduce or cancel the numbers beneath the square root signs. In this case they are all multiples of 5.

$$8\sqrt{5} - \sqrt{125} + \sqrt{80} = 8\sqrt{5} - \sqrt{25 \times 5} + \sqrt{16 \times 5}$$
$$= 8\sqrt{5} - 5\sqrt{5} + 4\sqrt{5}$$
$$= 7\sqrt{5}$$

Q Simplify $\dfrac{1}{2 + \sqrt{3}}$

The convention is to have no surds in the denominator, so aim to remove them. You can do this by multiplying.

Multiply top and bottom by $2 - \sqrt{3}$.

$$\frac{1}{2 + \sqrt{3}} = \frac{1}{2 + \sqrt{3}} \times \frac{2 - \sqrt{3}}{2 - \sqrt{3}} = \frac{2 - \sqrt{3}}{4 - 3}$$
$$= 2 - \sqrt{3}$$

This is called **rationalising the denominator**.

Division of polynomials

When you are faced with an expression such as $\frac{5x^3 + 18x^2 + 19x + 6}{5x + 3}$ you need to divide one polynomial by another.

In this case, you have a cubic divided by a linear expression. From the laws of indices, the result is a quadratic. There are two methods you could use to perform this division: either by long division, or by setting the quotient equal to a general polynomial of the required order.

Long division:

$$
\begin{array}{r}
x^2 + 3x + 2 \\
5x + 3 \overline{)\,5x^3 + 18x^2 + 19x + 6} \\
\underline{5x^3 + 3x^2} \qquad\qquad\quad \\
15x^2 + 19x \qquad \\
\underline{15x^2 + 9x} \qquad \\
10x + 6 \\
\underline{10x + 6} \\
0
\end{array}
$$

So $\frac{5x^3 + 18x^2 + 19x + 6}{5x + 3}$

$= x^2 + 3x + 2$

Setting the quotient equal to a polynomial of order 2:

$$\frac{5x^3 + 18x^2 + 19x + 6}{5x + 3} = ax^2 + bx + c$$

Comparing coefficients of:

x^3: $5 = 5a \Rightarrow a = 1$

x^2: $18 = 3a + 5b \Rightarrow b = 3$

x: $19 = 3b + 5c \Rightarrow c = 2$

Check constants: $6 = 3c$ ✓

So $\frac{5x^3 + 18x^2 + 19x + 6}{5x + 3}$

$= x^2 + 3x + 2$

Q Simplify $\frac{2x^3 - 5x^2 - 4x + 3}{x + 1}$

Long division:

$$
\begin{array}{r}
2x^2 - 7x + 3 \\
x + 1 \overline{)\,2x^3 - 5x^2 - 4x + 3} \\
\underline{2x^3 + 2x^2} \qquad\qquad\quad \\
-7x^2 - 4x \qquad \\
\underline{-7x^2 - 7x} \qquad \\
3x + 3 \\
\underline{3x + 3} \\
0
\end{array}
$$

So $\frac{2x^3 - 5x^2 - 4x + 3}{x + 1}$

$= 2x^2 - 7x + 3$

Setting the quotient equal to a polynomial of order 2:

$$\frac{2x^3 - 5x^2 - 4x + 3}{x + 1} = ax^2 + bx + c$$

Comparing coefficients of:

x^3: $2 = a \Rightarrow a = 2$

x^2: $-5 = a + b \Rightarrow b = -7$

x: $-4 = b + c \Rightarrow c = 3$

Check constants: $3 = c$ ✓

So $\frac{2x^3 - 5x^2 - 4x + 3}{x + 1} = 2x^2 - 7x + 3$

Algebraic fractions

To simplify expressions that include fractions you need to find the **lowest common denominator** and then express each term as a fraction in terms of that denominator.

Q Simplify $\frac{2}{x} - \frac{x}{x+2}$

The lowest common denominator is $x(x+2)$. Express each fraction in terms of $x(x+2)$.

$$\frac{2}{x} \times \frac{x+2}{x+2} - \frac{x}{x+2} \times \frac{x}{x} = \frac{2(x+2) - x^2}{x(x+2)} = \frac{4 + 2x - x^2}{x(x+2)}$$

Q Simplify $\frac{1}{1+x} - \frac{1}{1-x}$

The lowest common denominator is $(1+x)(1-x)$.

$$\frac{1}{1+x} \times \frac{1-x}{1-x} - \frac{1}{1-x} \times \frac{1+x}{1+x} = \frac{(1-x) - (1+x)}{(1+x)(1-x)} = \frac{-2x}{1-x^2} = \frac{2x}{x^2-1}$$

Q Simplify $\frac{2}{x^2+x} + \frac{3}{x^2-1}$

The **lowest** common denominator is **not** $(x^2+x)(x^2-1)$ as factorising the denominator of each term gives $x(x+1)$ and $(x+1)(x-1)$, and they have a common factor of $(x+1)$. The lowest common denominator is $x(x+1)(x-1)$.

$$\frac{2}{x^2+x} \times \frac{x(x+1)(x-1)}{x(x+1)(x-1)} + \frac{3}{x^2-1} \times \frac{x(x+1)(x-1)}{x(x+1)(x-1)}$$

$$= \frac{2}{x(x+1)} \times \frac{x(x+1)(x-1)}{x(x+1)(x-1)} + \frac{3}{(x+1)(x-1)} \times \frac{x(x+1)(x-1)}{x(x+1)(x-1)}$$

$$= \frac{2(x-1)}{x(x+1)(x-1)} + \frac{3x}{x(x+1)(x-1)}$$

$$= \frac{2x - 2 + 3x}{x(x+1)(x-1)}$$

$$= \frac{5x-2}{x(x^2-1)}$$

Simultaneous equations

Simultaneous equations are sets of equations in two (or more) variables, such as:

$$\begin{aligned} 13x + 9y &= 36 \\ 2x + 3y &= 6 \end{aligned} \qquad \text{or} \qquad \begin{aligned} 2x + 7y &= 42 \\ y &= x - 3 \\ x + 6y &= 25 \end{aligned}$$

These equations have a solution if there is one pair of values for x and y which satisfy *all* the equations in the set.

You can solve simultaneous equations graphically or algebraically.

Graphical solution of simultaneous equations

If you draw the graphs of the two equations, the solution is the point where the two lines intersect.

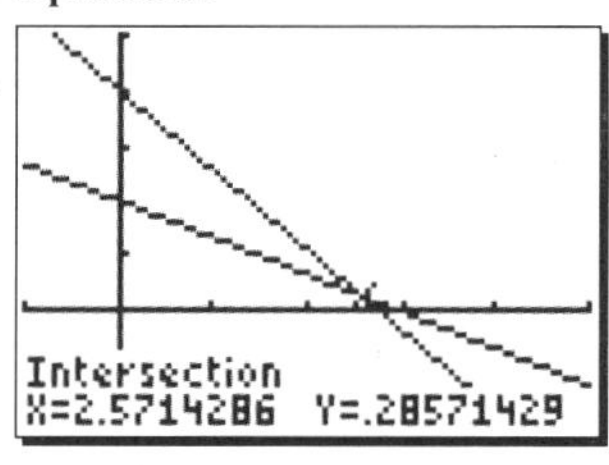

 Solve $13x + 9y = 36$
$2x + 3y = 6$

First of all, rearrange the two equations to make y the subject in both.

Use a graphics calculator to draw the graphs and find the intersection point.

So $x = 2.57$ and $y = 0.29$ is the solution to this set of equations.

Algebraic solution of simultaneous equations

There are two algebraic methods for solving simultaneous equations: the **substitution method** and the **elimination method**.

Substitution method

 Solve $2x + 3y = 30$ **(i)**
$x + y = 12$ **(ii)**

Rearrange **(ii)** to make x the subject of the equation:

$$x = 12 - y \quad \textbf{(iii)}$$

Now substitute this value into equation (i):

$$2(12 - y) + 3y = 30$$

Solve for y: $24 - 2y + 3y = 30 \Rightarrow y = 6$

Now find the corresponding value of x from equation (iii):

$$x = 12 - 6, x = 6$$

Therefore the solution is $x = 6$, $y = 6$.

Elimination method

Q Solve $2p + 3r = 11$ **(i)**
$3p + 4r = 15$ **(ii)**

You need to make the coefficient of p the same in both equations.

$3 \times$ (i): $6p + 9r = 33$ **(iii)**

$2 \times$ (ii): $6p + 8r = 30$ **(iv)**

(iii) – **(iv)** to eliminate p: $9r - 8r = 33 - 30$

$r = 3$

Use equation **(i)** to find p when $r = 3$: $2p + 3 \times 3 = 11$
$\Rightarrow 2p = 2 \Rightarrow p = 1$

Check, using the other equation: $3p + 4r = 15$
$\Rightarrow 3 \times 1 + 4 \times 3 = 3 + 12 = 15$ ✓

Therefore the solution to these equations is $p = 1$ and $r = 3$.

A quadratic and a linear equation

A straight line and a parabola (from a quadratic equation) may intersect at two, one or no points, as shown in the diagram.

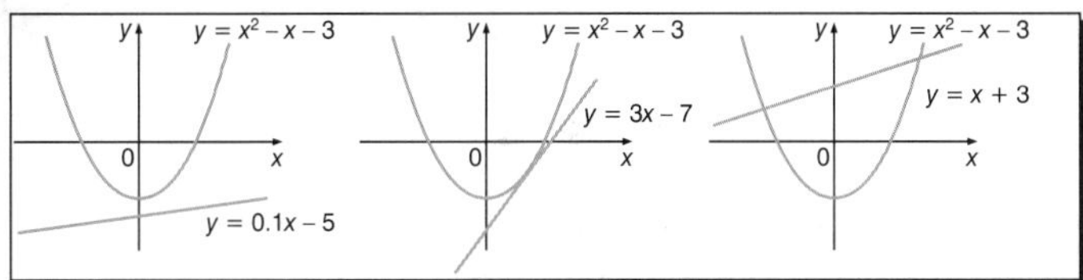

It is always worth sketching the graphs of the equations to see how many solutions there should be, but make sure that you have the whole picture.

Q Solve $y = 2x^2 - 3x - 1$ **(i)**
$y = 3x + 7$ **(ii)**

Using a graphics calculator shows that there are two solutions.

Substitute the value for y from **(ii)** into **(i)**: $3x + 7 = 2x^2 - 3x - 1$

Rearrange to form a quadratic: $2x^2 - 6x - 8 = 0$

Solve by factorising or quadratic formula: $x = -1, x = 4$

Find the corresponding values for y from **(ii)**: $x = -1, y = -3 + 7 = 4$
$x = 4, y = 3 \times 4 + 7 = 19$

Therefore the solutions are $(-1, 4)$ and $(4, 19)$.

Check yourself

Algebra

1 Solve

$$\frac{x}{2 - 3x} = \frac{2}{3}$$

2 Solve

$$\frac{x + 1}{2x - 3} = \frac{2}{3}$$

3 Simplify this expression.

$$\frac{1}{1 + x} - \frac{1}{1 - x}$$

4 Simplify this expression.

$$\frac{1}{2 - x} + \frac{1}{2 + x}$$

5 Simplify this expression.

$$\frac{2}{4 - x^2} + \frac{1}{x^2 + 2x}$$

6 Make l the subject of this formula.

$$T = 2\pi\sqrt{\frac{l}{g}}$$

7 Make I the subject of this formula.

$$H = I^2Rt$$

8 Make C the subject of this formula.

$$R = \tfrac{1}{2}dACv^2$$

9 Make m the subject of this formula.

$$f = \frac{1}{2l}\sqrt{\frac{T}{m}}$$

10 Solve this inequality.

$$2x - 5 > x + 2$$

11 Solve this inequality.

$$|4 - 3x| \geqslant 2$$

12 Solve this inequality.

$$6 - x - x^2 > 0$$

The answers are on page 95.

Check yourself

Algebra

13 Simplify these expressions.

(a) $\dfrac{2p^2q^5r^7}{(3pq^2r^3)^2}$ **(b)** $\sqrt[3]{2\frac{10}{27}}$

14 Without using a calculator find n if $2n^{-3} = \frac{1}{32}$

15 Simplify:

$2\sqrt{6} - \sqrt{150} + \sqrt{216}$

16 Simplify:

$\dfrac{3 - \sqrt{2}}{3 + \sqrt{2}}$

17 Simplify this quotient.

$\dfrac{2x^4 + 5x^3 - 23x^2 - 38x + 24}{x^2 - x - 6}$

18 Simplify this quotient.

$\dfrac{4x^4 - 2x^3 - 11x^2 - 5x}{x^2 + x}$

19 Solve this set of simultaneous equations.

$3x + 2y = 8$
$2x - y = 3$

20 Solve this set of simultaneous equations.

$3x + 2y = 6$
$x + 2y = 4$

21 Solve this set of simultaneous equations.

$y + 2 = 3x$
$x^2 - x - 2 = y$

The answers are on page 96.

Cartesian coordinates

You can describe a point in a plane uniquely by a pair of **Cartesian coordinates** (x, y) with respect to an origin O and two perpendicular axes Ox and Oy.

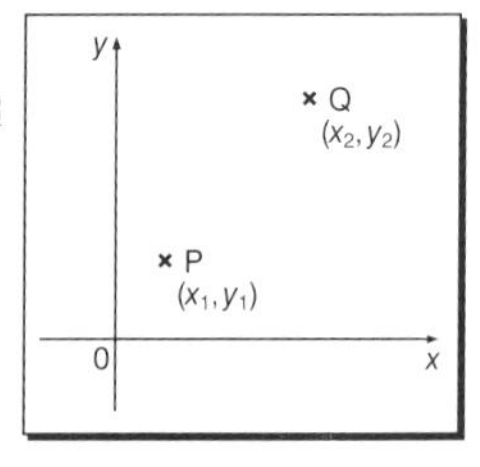

The distance d between the two points $P(x_1, y_1)$ and $Q(x_2, y_2)$ is given by

$d = \sqrt{(x_2 - x_1)^2 + (y_2 - y_1)^2}$.

The coordinates of the **midpoint** of the straight line joining P and Q are $(\frac{1}{2}(x_1 + x_2), \frac{1}{2}(y_1 + y_2))$.

The **gradient** of the straight line joining P and Q is $\dfrac{y_2 - y_1}{x_2 - x_1}$.

Two lines are **parallel** if they have the same gradient.

Two lines are **perpendicular** if the product of their gradients is –1.

The gradient of a line can be positive or negative. A line with positive gradient slopes up from left to right, a line with negative gradient slopes down from left to right. A zero gradient means a horizontal line.

Q The points A(–1, 3), B(4, 15) and C(8, 11) are in the Cartesian plane. Determine: **(a)** the distance between A and B **(b)** the coordinates of the midpoint of AB **(c)** the gradient of AB **(d)** the gradient of BC.

(a) Distance apart $= \sqrt{(4 - (-1))^2 + (15 - 3)^2}$
$= \sqrt{5^2 + 12^2} = \sqrt{169} = 13$ units

(b) The midpoint of AB has coordinates $(\frac{1}{2}(-1 + 4), \frac{1}{2}(3 + 15)) = (1.5, 9)$

(c) The gradient of AB $= \dfrac{15 - 3}{4 - -1} = \dfrac{12}{5} = 2.4$

(d) The gradient of BC $= \dfrac{11 - 15}{8 - 4} = \dfrac{-4}{4} = -1$

Q Show that **(a)** the line joining C(8, 11) to D(13, 23) is parallel to AB **(b)** the line joining C to E(10, 13) is perpendicular to BC.

(a) The gradient of CD $= \dfrac{23 - 11}{13 - 8} = \dfrac{12}{5} = 2.4$

This equals the gradient of AB, so AB and CD are parallel.

(b) The gradient of CE $= \dfrac{13 - 11}{10 - 8} = \dfrac{2}{2} = 1$

Since the gradient of BC = –1, the product of the two is –1 and hence CE is perpendicular to BC.

Equation of a straight line

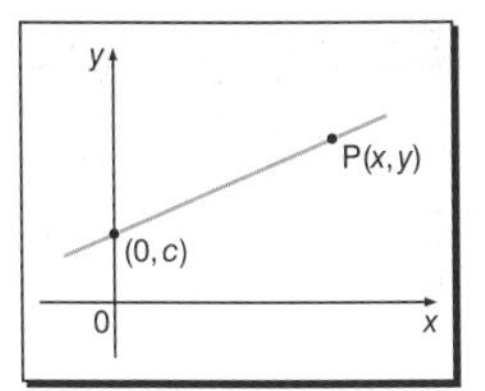

The straight line shown in the diagram has gradient (or slope) m and passes through the point $(0, c)$. Let $P(x, y)$ be a general point on the line. The gradient, $m = \dfrac{y - c}{x - 0}$ so $y = mx + c$.

This is the **general equation** of a straight line with gradient m and y-intercept c.

The equation of a straight line with slope m passing through the point (a, b) is given by $y - b = m(x - a)$.

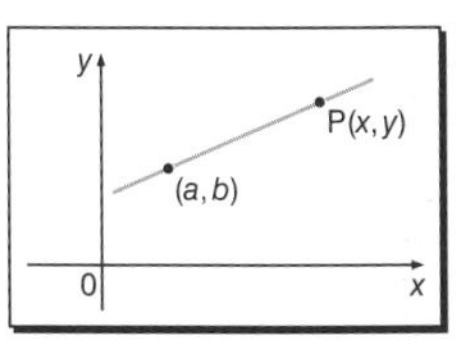

The equation of a straight line through two points (x_1, y_1) and (x_2, y_2) is given by $y - y_1 = \dfrac{y_2 - y_1}{x_2 - x_1}(x - x_1)$.

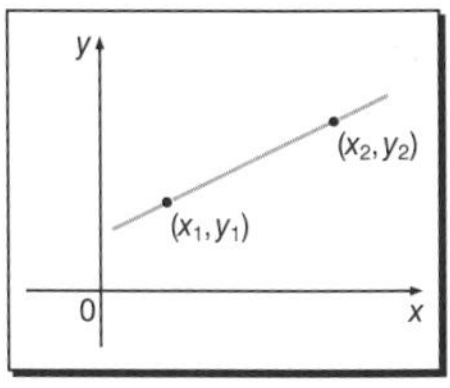

Q **(a)** Find the equation of the straight line with gradient –2 passing through (1, 3). Where does it intercept the x- and y-axes? Sketch the line.

(b) Find the equation of the straight line joining (4, 15) and (8, 3).

(c) Find the equation of the line passing through (4, 15), perpendicular to the line found in (b).

(a) $y - 3 = -2(x - 1) \Rightarrow y = -2x + 5$

The line intersects the x-axis when $y = 0 \Rightarrow x = 2.5$.

The y-intercept $= 5$.

(b) $y - 15 = \dfrac{15 - 3}{4 - 8}(x - 4)$

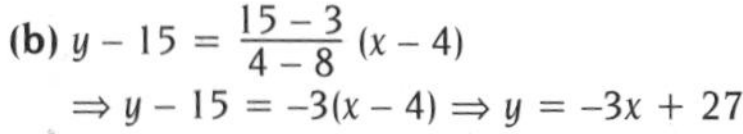

$\Rightarrow y - 15 = -3(x - 4) \Rightarrow y = -3x + 27$

(c) The line in (**b**) has gradient –3 so the gradient of the line perpendicular to it is $\frac{1}{3}$.

Therefore its equation is $y - 15 = \frac{1}{3}(x - 4) \Rightarrow 3y - x = 41$.

Fitting straight lines to data

Some sets of experimental data can be represented quite well by a straight line, even if the line does not pass through every data point. The **line of best fit** is the straight line that best balances out the over and under estimates. You can find this line by using a graphics calculator or by eye, using your judgement. You can put an additional constraint on the line, because it must pass through the point $M(\bar{x}, \bar{y})$ i.e. the mean values of the x- and y-coordinates.

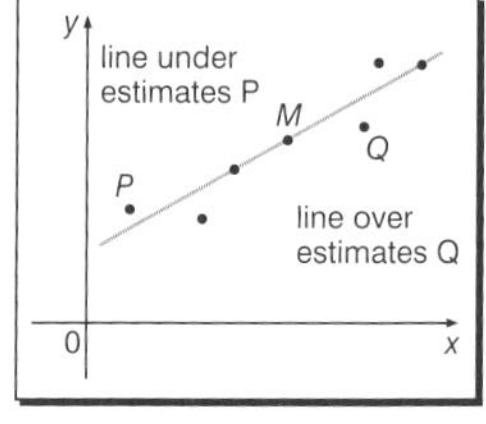

Find the line of best fit by plotting M, then pivoting a straight edge about it until the over and under estimates balance out. Then calculate the gradient and y-intercept of the line from the graph to find the equation.

Q The table shows temperatures on the Celsius (°C) and Fahrenheit (°F) scales. Plot a graph of Fahrenheit temperatures against Celsius. Confirm that there is a linear relationship and find the equation.

	°C	°F
Freezing point of water	0	32
Boiling point of water	100	212
Human body temperature	37	98
Summer day	20	68
Washing-up water	50	122
$M(\bar{x}, \bar{y}) = (41.4, 106.4)$		

The graph shows that the relationship is linear as all the data points lie on a straight line through $(\bar{x}, \bar{y})$. The gradient is $\dfrac{212 - 32}{100 - 0}$ and the intercept on the vertical axis is 32. The relationship is therefore given by $F = 1.8C + 32$.

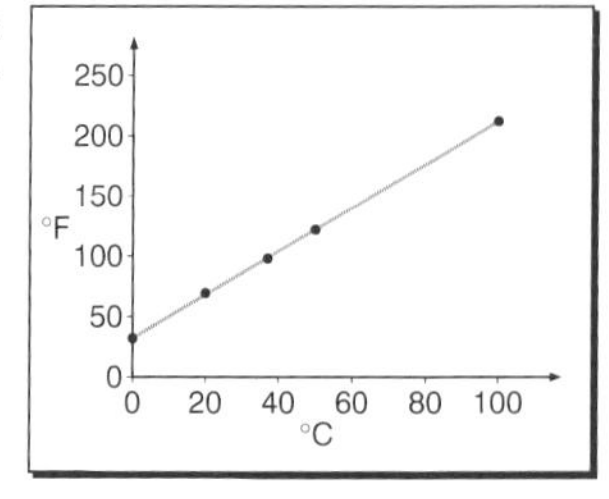

This equation can be used to find the Fahrenheit equivalent of any Celsius value in the range [0, 100] by **interpolation**.

A line of best fit can't be used for values outside the range of given data (**extrapolation**) because there is no guarantee that the linear relationship will hold (although the relationship above *is* true for all temperatures).

Check yourself

Linear Functions

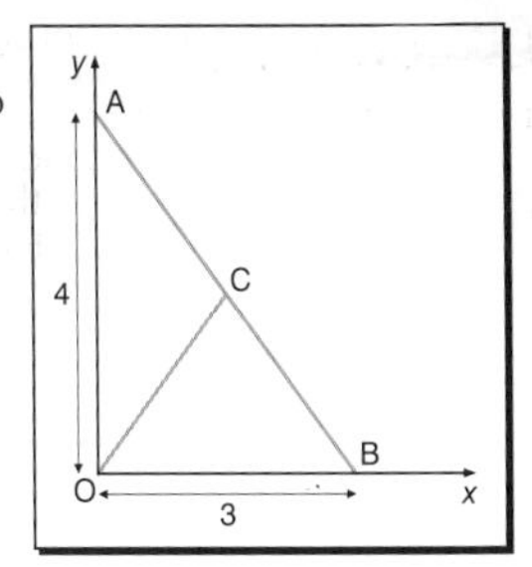

1 The diagram shows a framework used to support a horizontal platform OB attached to a vertical wall OA. Use the axes shown to:

(a) write down the coordinates of the points A and B

(b) write down the coordinates of C, the midpoint of AB

(c) find the length of AB

(d) find the equations of the lines AB and OC.

2 Find the gradient and the equation of the straight line that passes through the two points A(1, 4) and B(3, –2). Find also the equation of the straight line through B, perpendicular to AB.

3 The data in the table shows the stretched length l (m) of a vertically suspended spring when different masses M (kg) are attached to its free end.

Load attached M (kg)	10.0	17.5	21.5	25.5	29.0
Length of spring l (m)	0.4	0.585	0.68	0.77	0.87

Confirm that the data follows a linear relationship of the form $l = aM + b$ and find the values of a and b. What is the length of the unstretched spring? What mass would be required to give a stretched length of 0.8 m?

4 A young sprinter was looking back over the best times she had recorded in 100 m races each year. The times are shown in the table.

Age (years)	12	13	14	15	16	17	18	19
Time (seconds)	19.1	15.5	15.1	14.6	12.2	12.1	11.3	10.8

Investigate the relationship between the sprinter's age and her best times. Use the equation you calculate to predict the time taken for 100 m when she is **(a)** 20 years old **(b)** 30 years old. Comment.

The answers are on page 97.

Quadratic equations

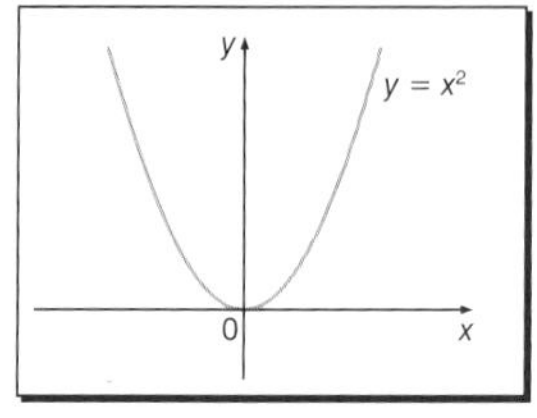

The general form of the quadratic expression is $ax^2 + bx + c$ where a, b and c may take any values ($a \neq 0$). The shape of the graph of the quadratic equation $y = ax^2 + bx + c$ is called a **parabola**.

An equivalent form is $a(x + p)^2 + q$ where p and q can be positive or negative. If $a > 0$ then the curve is u-shaped, if $a < 0$ then the curve is n-shaped. The basic parabola has the symmetric shape as shown. The shape is affected by the value of a. The graph of $a(x + p)^2 + q$ is obtained from:

(i) a horizontal shift of p units (to the left if $p > 0$, to the right if $p < 0$) and

(ii) a vertical shift of q units (upwards if $q > 0$, downwards if $q < 0$).

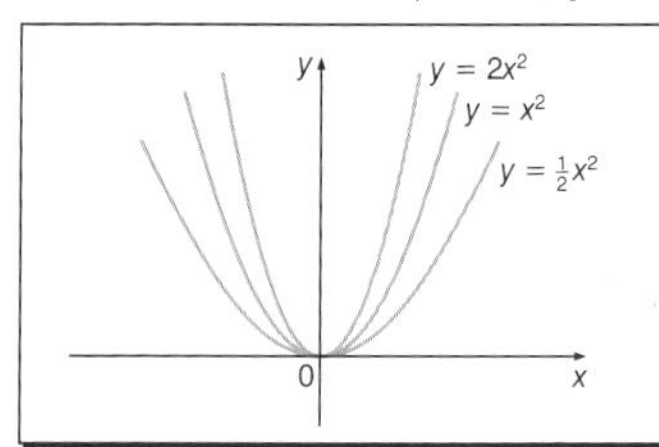

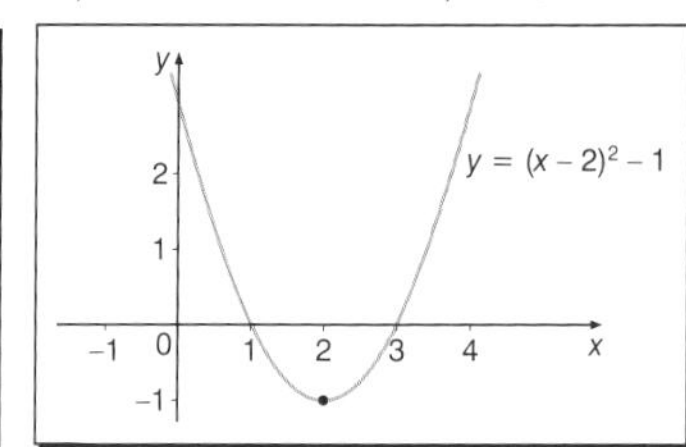

Expanding brackets

Expanding brackets means multiplying them out.

$(x + a)(x + b) = x^2 + (a + b)x + ab$

You may have your own method for doing this.

Q Expand these brackets.

(a) $(x - 1)(x + 2)$ **(b)** $(x + 3)(x - 2)$ **(c)** $(x - 2)(x + 2)$

(d) $(3 - x)(2 + x)$ **(e)** $(2x + 1)(x - 3)$

(a) $(x - 1)(x + 2) = x^2 + 2x - x - 2 = x^2 + x - 2$

(b) $(x + 3)(x - 2) = x^2 - 2x + 3x - 6 = x^2 + x - 6$

(c) $(x - 2)(x + 2) = x^2 + 2x - 2x - 4 = x^2 - 4$

There is no middle term when $a = -b$.

This is called the **difference of two squares**.

(d) $(3 - x)(2 + x) = 6 + 3x - 2x - x^2 = 6 + x - x^2$

(e) $(2x + 1)(x - 3) = 2x^2 - 6x + x - 3 = 2x^2 - 5x - 3$

Factorising expressions

Factorisation is the mathematical opposite or inverse of expansion. This means that you start with an expression of the form $x^2 + 7x + 12$ and find the equivalent form $(x + 3)(x + 4)$. The two terms $(x + 3)$ and $(x + 4)$ are the **factors** of the quadratic. Since $(x + a)(x + b) = x^2 + (a + b)x + ab$ you need to find numbers (a and b) such that $ab = 12$ and $a + b = 7$.

Not all quadratics can be factorised.

Q Factorise the expression $x^2 - 9x + 18$.

Comparing $x^2 - 9x + 18$ with $x^2 + (a + b)x + ab$ you should find that $ab = 18$ and $(a + b) = -9$.

If you cannot see values for a and b straight away, then one method is to list all the possible pairs (a, b) which satisfy $ab = 18$:
$(1, 18)$, $(-1, -18)$, $(2, 9)$, $(-2, -9)$, $(6, 3)$ and $(-6, -3)$.
Inspect these pairs to find which one also satisfies $a + b = -9$. The only possibility is $(-6, -3)$ and hence $x^2 - 9x + 18 = (x - 6)(x - 3)$.

Difference of two squares

The **difference of two squares** describes expressions of the form $x^2 - a^2$. Factorising $x^2 - a^2$ gives $x^2 - a^2 = (x - a)(x + a)$. Expressions such as $x^2 - 4$, $x^2 - 9$, $x^2 - 16$ are differences of two squares.

Q Simplify these expressions.
(a) $x^2 - 25$ **(b)** $(x - 7)(x + 7)$ **(c)** $x^2 - 1$

(a) $x^2 - 25 = x^2 - 5^2 = (x - 5)(x + 5)$
(b) $(x - 7)(x + 7) = x^2 - 7^2 = x^2 - 49$
(c) $x^2 - 1 = x^2 - 1^2 = (x - 1)(x + 1)$

The next example shows how factorisation and the difference of two squares can be used to simplify rational algebraic expressions.

Q Simplify:
(a) $\dfrac{x^2 - x - 6}{x^2 - 9}$ **(b)** $\dfrac{x^2 - 4}{x^2 - 3x - 10}$

(a) $$\frac{x^2 - x - 6}{x^2 - 9} = \frac{(x - 3)(x + 2)}{(x - 3)(x + 3)} = \frac{x + 2}{x + 3}$$

(b) $$\frac{x^2 - 4}{x^2 - 3x - 10} = \frac{(x - 2)(x + 2)}{(x + 2)(x - 5)} = \frac{x - 2}{x - 5}$$

Solving quadratic equations

The equation $x^2 + 5x + 6 = 0$ can be factorised to give $(x + 2)(x + 3) = 0$. The values $x = -3$ and $x = -2$ **satisfy** the quadratic equation $x^2 + 5x + 6 = 0$. The values of -2 and -3 are the **roots** of the equation. The two roots $x = -2$ and $x = -3$ are **rational roots** as they are integers or fractions. The roots of a quadratic such as $x^2 = 20$ are $x = \pm\sqrt{20} = \pm 2\sqrt{5}$. These are **irrational roots** as they cannot be expressed exactly as an integer or fraction. If $(x - a)$ is a factor of $f(x)$, $x = a$ is a root of the equation $f(x) = 0$ and $f(a) = 0$.

You may need to simplify an equation, before it reveals itself as a quadratic.

Q Solve the equation $x - \frac{18}{x} = 7$.

First multiply through by x to clear the fractions: $x^2 - 18 = 7x$

Then rearrange: $x^2 - 7x - 18 = 0$

Now factorise: $x^2 - 7x - 18 = (x + a)(x + b)$

$\Rightarrow a + b = -7$ and $ab = -18 \Rightarrow a = -9$ and $b = 2$.

$(x - 9)(x + 2) = 0$ and $x = 9$ and $x = -2$ are the two solutions.

Q The product of two consecutive positive odd numbers is 575. Find the numbers.

An odd number can be represented by the expression $(2n - 1)$, $n = 1, 2, 3, \ldots$. The next odd number after $(2n - 1)$ is $(2n - 1) + 2 = 2n + 1$. The statement in the question is equivalent to $(2n - 1)(2n + 1) = 575$. This is a difference of two squares.

$(2n)^2 - 1^2 = 575 \Rightarrow 4n^2 = 576 \Rightarrow n^2 = 144 \Rightarrow n = \pm 12$

Then the two odd numbers are $2 \times 12 - 1 = 23$ and $2 \times 12 + 1 = 25$ or $2 \times -12 - 1 = -25$ and $2 \times -12 + 1 = -23$. The question asks for positive numbers, so the solutions are $x = 23$, and $x = 25$.
(**Check:** $23 \times 25 = 575$)

Q By letting $y = 2^x$, solve the equation $2^{2x+1} - 5 \times 2^x + 2 = 0$.

Use the laws of indices to rewrite 2^{2x+1} as $2^x \times 2^x \times 2 = 2y^2$. Then the original equation is equivalent to the quadratic $2y^2 - 5y + 2 = 0$. Factorisation is of the form $(2y + a)(y + b) \Rightarrow a + 2b = -5$ and $ab = 2$. Considering possible values for (a, b) gives $a = -1$ and $b = -2$.

$(2y - 1)(y - 2) = 0 \Rightarrow y = \frac{1}{2}$ and $y = 2$.

But $y = 2^x \Rightarrow 2^x = \frac{1}{2} \Rightarrow x = -1$ and $2^x = 2 \Rightarrow x = 1$.

The two solutions are $x = -1$ and $x = 1$.

Completing the square

You can rearrange a quadratic of the form $ax^2 + bx + c$ so that it involves a perfect square: $a(x + p)^2 + q$. This is called **completing the square**.

Q Use the method of completing the square to express $x^2 + 6x - 6$ in the form $a(x + p)^2 + q$ and hence solve $x^2 + 6x - 6 = 0$.

Expanding $a(x + p)^2 + q$ gives $ax^2 + 2apx + ap^2 + q$. Compare the expansion and $x^2 + 6x - 6$ to find a, p and q.

Compare coefficients: x^2: $a = 1$; x: $2ap = 6 \Rightarrow p = 3$

Compare constants: $ap^2 + q = -6 \Rightarrow q = -15$

So $x^2 + 6x - 6 = 0$ is equivalent to $(x + 3)^2 - 15 = 0$ and can be solved.
$(x + 3)^2 = 15 \Rightarrow x + 3 = \pm\sqrt{15} \Rightarrow x = -3 + \sqrt{15}$ and $x = -3 - \sqrt{15}$.

Q Rearrange the expression $20 + x - x^2$ by completing the square.

Compare coefficients in $20 + x - x^2$ and $ax^2 + 2apx + ap^2 + q$.

x^2: $a = -1$; x: $2ap = 1 \Rightarrow p = -\frac{1}{2}$

Compare constants: $ap^2 + q = 20 \Rightarrow q = 20 + \frac{1}{4} = \frac{81}{4}$

Hence $20 + x - x^2 = \frac{81}{4} - (x - \frac{1}{2})^2$

The quadratic formula

Completing the square on the general quadratic gives:

$$ax^2 + bx + c = a\left(x + \frac{b}{2a}\right)^2 - \frac{b^2 - 4ac}{4a}$$

Using this expression to solve $ax^2 + bx + c = 0$ leads to the quadratic formula: $x = \dfrac{-b \pm\sqrt{b^2 - 4ac}}{2a}$

The quantity $b^2 - 4ac$ is the **discriminant**.

If $b^2 - 4ac > 0$ the quadratic has two different roots.

If $b^2 - 4ac = 0$ the quadratic has one repeated root.

If $b^2 - 4ac < 0$ the quadratic has no roots.

Q Use the quadratic formula to solve these equations.
(a) $x^2 - 11x + 10 = 0$ **(b)** $x^2 + 4x + 5 = 0$

(a) $a = 1$, $b = -11$, $c = 10$: $x = \dfrac{11 \pm\sqrt{(-11)^2 - 4 \times 1 \times 10}}{2 \times 1} = 10, 1$.

(b) $a = 1$, $b = 4$, $c = 5$: $x = \dfrac{-4 \pm\sqrt{4^2 - 4 \times 1 \times 5}}{2 \times 1} = \dfrac{-4 \pm\sqrt{-4}}{2}$

As the discriminant is negative, there are no roots.

Check yourself

Quadratic Functions

1 Express $2x - 3 - x^2$ in the form $b - (x + a)^2$.
Show how this form can be used to sketch the graph of $y = 2x - 3 - x^2$ starting from the basic quadratic curve $y = -x^2$.

2 Expand the brackets and simplify the expression $(2x - 1)(x + 3) - (2x + 1)^2$.

3 Factorise the quadratic expression $2x^2 - 7x - 4$ and hence find the roots of the equation $2x^2 - 7x - 4 = 0$.

4 Use factorisation and the difference of two squares to simplify the rational expression $\dfrac{2x^2 + 7x + 6}{x^2 - 4}$.

5 Solve the equation $x = 1 + \dfrac{1}{x}$.

6 Use the method of completing the square on the quadratic function $y = 2 - x - x^2$ to express it in the form $b - (x + a)^2$. Hence sketch its graph.
Use the graph to confirm that the equation $y = 2 - x - x^2$ has two roots. Use the quadratic formula to find the values of the two roots.

The answers are on page 99.

Polynomial functions

In a **polynomial function** every term is a multiple of an integer power of x. The **degree** is the index of the highest power of x. $2x^3 - x^2 + 6x - 3$ is a polynomial of degree 3 (a **cubic**). You can use addition, multiplication, subtraction and division to combine polynomials.

Q Find the product of $2x^2 + x + 3$ and $x^3 - x^2 + 1$.

Multiply each term of $2x^2 + x + 3$ by each term of $x^3 - x^2 + 1$.

$$\begin{aligned}&(2x^2 + x + 3)(x^3 - x^2 + 1)\\&= 2x^2(x^3 - x^2 + 1) + x(x^3 - x^2 + 1) + 3(x^3 - x^2 + 1)\\&= 2x^5 - 2x^4 + 2x^2 + x^4 - x^3 + x + 3x^3 - 3x^2 + 3\\&= 2x^5 - x^4 + 2x^3 - x^2 + x + 3\end{aligned}$$

Q Simplify the quotient $\dfrac{2x^4 + 5x^3 - 11x^2 - 20x + 12}{x^2 + x - 6}$

From the laws of indices a polynomial of degree 4 divided by one of degree 2 gives a polynomial of degree 2 of the form $ax^2 + bx + c$. So $2x^4 + 5x^3 - 11x^2 - 20x + 12 = (x^2 + x - 6)(ax^2 + bx + c)$.

Compare coefficients of different powers of x to find a , b and c.

x^4: $2 = a$; x^3: $5 = a + b \Rightarrow b = 3$; constants: $12 = -6c \Rightarrow c = -2$

$$\frac{2x^4 + 5x^3 - 11x^2 - 20x + 12}{x^2 + x - 6} = 2x^2 + 3x - 2$$

Roots of an equation and the factor theorem

$x = a$ is a **root** of the equation $f(x) = 0$ if $f(a) = 0$. The graph of $y = f(x)$ crosses the x-axis at $x = a$, so $(x - a)$ is a factor of $f(x)$. This result is summarised in the **factor theorem**:

For any polynomial function $f(x)$, $f(a) = 0 \Leftrightarrow (x - a)$ is a factor.

A polynomial of degree n has at most n roots.

Q $f(x) = 2x^3 + 3x^2 - 3x - 2$. Show that $(x - 1)$ is a factor of $f(x)$. Express $f(x)$ as the product of a linear and quadratic factor and hence solve the equation $f(x) = 0$ as fully as possible.

From the factor theorem, if $(x - 1)$ is a factor then $2 \times 1^3 + 3 \times 1^2 - 3 \times 1 - 2$ should be zero. It is, so $(x - 1)$ is a factor.

So $2x^3 + 3x^2 - 3x - 2 = (x - 1)(ax^2 + bx + c)$.
Comparing coefficients gives $a = 2$, $b = 5$ and $c = 2$.

So $2x^3 + 3x^2 - 3x - 2 = (x - 1)(2x^2 + 5x + 2) = (x - 1)(2x + 1)(x + 2)$.

Then $f(x) = 0$ for $x = 1$, $x = -\frac{1}{2}$ and $x = -2$.

Polynomial equations and inequalities

Suppose you are given a polynomial inequality of the form $x^4 - 7x^2 + 6 \leqslant 0$. You need to find the range(s) of x which satisfy the inequality. Using a graphical method, you could produce the graph of $y = x^4 - 7x^2 + 6$ and then simply identify the ranges of x for which $y \leqslant 0$. Alternatively, an algebraic approach is to find the roots of $f(x) = x^4 - 7x^2 + 6 = 0$ and investigate the sign of each factor on each inter-root interval. Then you can find the sign of $f(x)$ for each interval.

Q **(a)** Show that $x = -1$ and $x = 1$ are roots of the equation $x^4 - 7x^2 + 6 = 0$ and use the factor theorem to write down two factors of $f(x) = x^4 - 7x^2 + 6$.

(b) Factorise $f(x)$ completely.

(c) Hence find the regions for which $x^4 - 7x^2 + 6 \leqslant 0$.

(d) Sketch the graph of $y = f(x)$.

(a) $f(-1) = 0$ so $x = -1$ is a root and $(x + 1)$ is a factor.
$f(1) = 0$ so $x = 1$ is a root and $(x - 1)$ is a factor.

(b) You can represent the polynomial of degree 4 by the product of two linear factors $(x + 1)$ and $(x - 1)$, and a quadratic factor.

$$x^4 - 7x^2 + 6 = (x + 1)(x - 1)(ax^2 + bx + c)$$
$$x^4 - 7x^2 + 6 = (x^2 - 1)(ax^2 + bx + c)$$
$$= ax^4 + bx^3 + (c - 1)x^2 - bx - c$$

Comparing coefficients on both sides gives $a = 1$, $b = 0$ and $c = -6$.
Hence $x^4 - 7x^2 + 6 = (x + 1)(x - 1)(x^2 - 6)$
$= (x + 1)(x - 1)(x + \sqrt{6})(x - \sqrt{6})$

Thus $x^4 - 7x^2 + 6 = 0$ for $x = 1, -1, \sqrt{6}, -\sqrt{6}$.
Now investigating the sign in each region:

	-3	$-\sqrt{6}$	-2	-1	0	1	2	$\sqrt{6}$	3
$(x - 1)$	⊖		⊖		⊖		⊕		⊕
$(x + 1)$	⊖		⊖		⊕		⊕		⊕
$(x - \sqrt{6})$	⊖		⊖		⊖		⊖		⊕
$(x + \sqrt{6})$	⊖		⊕		⊕		⊕		⊕
$f(x)$	⊕		⊖		⊕		⊖		⊕

It is apparent that $f(x) \leqslant 0$ for the two intervals
$-\sqrt{6} \leqslant x \leqslant -1$ and $1 \leqslant x \leqslant \sqrt{6}$.

(d) This information will also help to sketch the graph of $y = f(x)$.

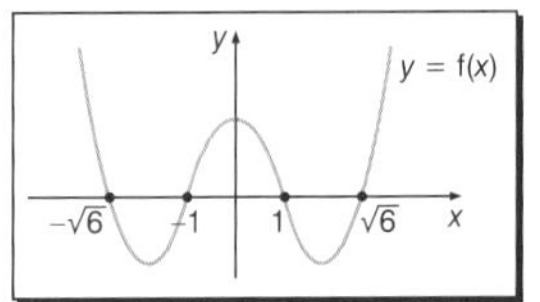

Check yourself

Cubic and Polynomial Functions

1 Simplify this quotient.

$$\frac{2x^4 + 5x^3 - 23x^2 - 38x + 24}{6 + x - x^2}$$

2 **(a)** Show that $x = -1$ is a root of the cubic equation $x^3 - 3x^2 + 4 = 0$.

(b) Hence write down a factor of $f(x) = x^3 - 3x^2 + 4$.

(c) Factorise $f(x)$ completely and describe the roots fully.

(d) Solve the inequality $f(x) > 0$.

3 The function f is defined by $f(x) = x^3 - 7x - 6$.

(a) Use the factor theorem to show that $(x - 3)$ is a factor of $f(x)$.

(b) Write $f(x)$ in the form
$f(x) = (x - 3)(ax^2 + bx + c)$
giving the values of a, b and c.

(c) Hence solve $f(x) = 0$.

(d) Use your solutions to $f(x) = 0$ to **write down** the solutions to the equation $f(x + 1) = 0$.

The answers are on page 100.

Differentiation

Differentiation is the mathematical process of finding the **rate of change** of a function f(x). Graphically it may be interpreted as the **slope** or **gradient** of the tangent, at a point on the graph of the function.

The result of differentiating a function $y = f(x)$ is the **derivative** of the function and is denoted by $\frac{dy}{dx}$ or $f'(x)$.

The formal definition of the derivative is $f'(x) = \lim_{h \to 0} \frac{f(x + h) - f(x)}{h}$.

You can use this result to establish that if $f(x) = x^n$ then $f'(x) = nx^{n-1}$ where n is an integer. The derivative of a **constant** is zero because constants do not change.

The derivative of a sum (or difference) of terms is equal to the sum (or difference) of the derivatives of the terms, so you can differentiate polynomials.

Q If $f(x) = x^4 + 2x^3 + 3x^2 + 4x + 5$ find $f'(x)$.

If $f(x) = x^4 + 2x^3 + 3x^2 + 4x + 5$ then

$$f'(x) = 4x^3 + 2(3x^2) + 3(2x) + 4(1) + 0$$
$$= 4x^3 + 6x^2 + 6x + 4$$

Stationary points

The **stationary value** of a function is the value of the function at a point where its derivative is zero. Since the derivative can be interpreted as the slope of the tangent, then stationary values correspond to those points where the tangent has zero slope, which is where the tangent is horizontal.

In the diagram, P and R are local maxima, Q is a local minimum and S is a point of inflexion. The word 'local' is included in the description to emphasise that although the graph may have other points where the value of the function is numerically greater (or smaller), the key characteristic of a local maximum (or minimum) is that it is a point where the tangent is horizontal.

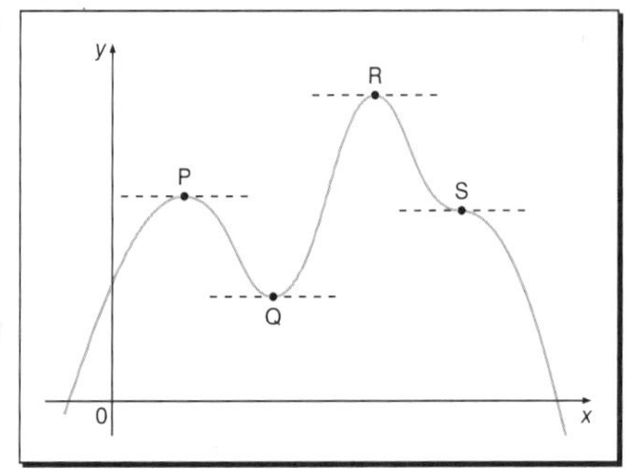

To distinguish the three types of stationary point, examine the behaviour of f'(x) either side of (and close to) the stationary point.

Q Show that the function $f(x) = x^5(x - 3)$ has a point of inflexion at (0, 0) and a local minimum at (2.5, –48.83).

First expand f(x) to give $f(x) = x^6 - 3x^5$ and then differentiate it to give $f'(x) = 6x^5 - 15x^4$. The stationary points are found by solving $6x^5 - 15x^4 = 0 \Rightarrow 3x^4(2x - 5) = 0 \Rightarrow x = 0, 2.5$. Substituting these values into f(x) gives the coordinates of the stationary points as (0, 0) and (2.5, –48.83). To classify them, consider f'(x) either side of the stationary point.

Consider the point (0, 0). Examine f'(x) at $x = \pm 0.1$:
$f'(-0.1) < 0$ and $f'(0.1) < 0 \Rightarrow (0, 0)$ is a point of inflexion as the gradient has the same sign immediately to the left and right of the stationary point.

Similarly for the point (2.5, –48.83): $f'(2.4) < 0$ and $f'(2.6) > 0 \Rightarrow (2.5, -48.83)$ is a local minimum as the gradient changes from negative to positive on passing through the stationary point. This can be verified by using a graphics calculator to sketch the function.

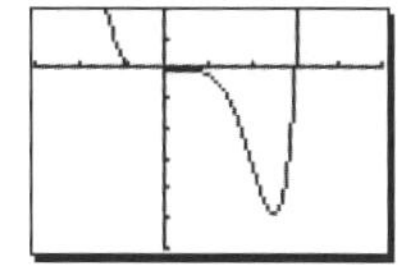

Optimisation

Since the solution(s) of $f'(x) = 0$ can correspond to local maxima and minima, you can use this method in problems where the objective is to maximise (or minimise) a function. This process is called **optimisation**.

Q The deflection y at any point on a beam, of length 2 m, clamped horizontally at one end and resting on a support at the same level at the other end, is given by $y = f(x) = -2x^4 + 10x^3 - 12x^2$. Show that y is a minimum when $x = 1.157$ m.

For maxima or minima $f'(x) = 0 \Rightarrow 8x^3 - 30x^2 + 24x = 0$. $x(8x^2 - 30x + 24) = 0 \Rightarrow x = 0$ or $x = 1.157, 2.593$ (using the quadratic formula). Discard the result 2.593 as it lies beyond the end of the beam.

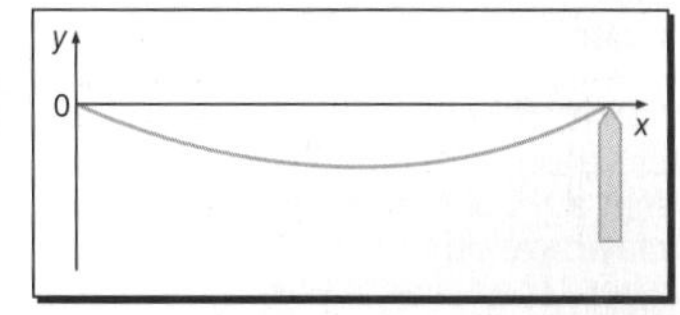

To show that y is a minimum when $x = 1.157$ examine f'(1.15) and f'(1.16). $f'(1.15) < 0$ and $f'(1.16) > 0$. Taken together with $f'(1.157) = 0$, this confirms that $x = 1.157$ m corresponds to minimum deflection.

Differentiation of functions with rational or negative indices

You may need to differentiate expressions such as $x^{\frac{3}{2}}$, x^{-2}, $\sqrt{x} = x^{\frac{1}{2}}$, $\frac{1}{\sqrt{x}} = x^{-\frac{1}{2}}$, $\frac{1}{\sqrt[3]{x}} = x^{-\frac{1}{3}}$. The rule that if $f(x) = x^n$ then $f'(x) = nx^{n-1}$ stated earlier for integer n also applies if n is rational or negative.

Q Differentiate $f(x) = \sqrt{x} + \frac{1}{\sqrt{x}} + x^3 + \frac{1}{x^2} + \sqrt[3]{x}$ with respect to x.

First express each term in the form x^k for suitable k to give:

$f(x) = x^{\frac{1}{2}} + x^{-\frac{1}{2}} + x^3 + x^{-2} + x^{\frac{1}{3}}$

Differentiate each term, using the rule quoted above, to give

$f'(x) = \frac{1}{2}x^{\frac{1}{2}-1} - \frac{1}{2}x^{-\frac{1}{2}-1} + 3x^2 - 2x^{-2-1} + \frac{1}{3}x^{\frac{1}{3}-1}$

$$f'(x) = \frac{1}{2}x^{-\frac{1}{2}} - \frac{1}{2}x^{-\frac{3}{2}} + 3x^2 - 2x^{-3} + \frac{1}{3}x^{-\frac{2}{3}}$$

$$= \frac{1}{2x^{\frac{1}{2}}} - \frac{1}{2x^{\frac{3}{2}}} + 3x^2 - \frac{2}{x^3} + \frac{1}{3x^{\frac{2}{3}}}$$

$$= \frac{1}{2\sqrt{x}} - \frac{1}{2\sqrt{x^3}} + 3x^2 - \frac{2}{x^3} + \frac{1}{3\sqrt[3]{x^2}}$$

Tangents and normals

You can use the fact that $f'(x)$ can be interpreted as the slope of the tangent to the curve $y = f(x)$ to find the equation of the tangent at any point. Since the product of the slopes of two perpendicular lines is equal to -1, you can also find the slope of the normal and hence its equation.

Q Find the equation of the tangent and normal to the curve $y = x + \frac{1}{x}$ at the point (2, 2.5).

$$y = x + \frac{1}{x} = x + x^{-1} \Rightarrow \frac{dy}{dx} = 1 - x^{-2} = 1 - \frac{1}{x^2}$$

When $x = 2$ the slope of the tangent $= 1 - \frac{1}{2^2} = \frac{3}{4}$.

The equation of the tangent is $y - 2.5 = \frac{3}{4}(x - 2) \Rightarrow 4y - 3x = 4$.

The slope of the normal at (2, 2.5) is $-\frac{4}{3}$ (because $\frac{3}{4} \times -\frac{4}{3} = -1$).

Its equation is $y - 2.5 = -\frac{4}{3}(x - 2) \Rightarrow 3y + 4x = 15.5$.

Differentiation of exponential and logarithmic functions

If $f(x) = e^{kx}$ (k constant) then $f'(x) = ke^{kx}$.

If $f(x) = \ln x$ then $f'(x) = \dfrac{1}{x}$.

Q Show that the curve $y = f(x) = e^{0.5x} - 2x + 1$ has a stationary point when $x = \ln 16$. Use your graphic calculator to sketch the graph and show that it is a minimum point.

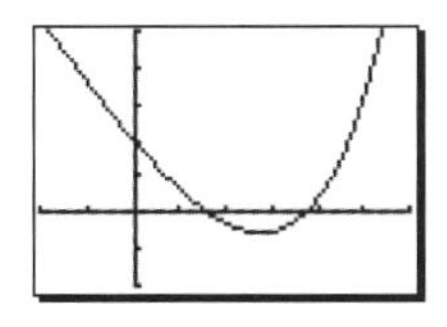

$f(x) = e^{0.5x} - 2x + 1 \Rightarrow f'(x) = 0.5e^{0.5x} - 2$.
Thus at a stationary point
$0.5e^{0.5x} - 2 = 0 \Rightarrow e^{0.5x} = 4$.
To solve this equation for x take natural logarithms of both sides to give $0.5x = \ln 4$ i.e.
$x = 2\ln 4 = \ln 4^2 = \ln 16$.
The graphics calculator display shows that this is a minimum.

Q Show that the curve $y = f(x) = 3x - x^2 - \ln x$ has stationary points at $(1, 2)$ and $(\frac{1}{2}, \frac{5}{4} + \ln 2)$. Confirm that $(1, 2)$ is a local maximum.

$f(x) = 3x - x^2 - \ln x \Rightarrow f'(x) = 3 - 2x - \dfrac{1}{x}$

For stationary points $3 - 2x - \dfrac{1}{x} = 0 \Rightarrow (2x - 1)(x - 1) = 0 \Rightarrow x = \frac{1}{2}, 1$.

When $x = \frac{1}{2}$, $f(\frac{1}{2}) = \frac{3}{2} - \frac{1}{4} - \ln\frac{1}{2} = \frac{5}{4} + \ln 2$

and when $x = 1$, $f(1) = 3 - 1 - \ln 1 = 2$.

The stationary points are at $(1, 2)$ and $(\frac{1}{2}, \frac{5}{4} + \ln 2)$.

To classify the point at $(1, 2)$, examine the signs of $f'(0.9)$ and $f'(1.1)$.

$f'(0.9) = 0.089$ and $f'(1.1) = -0.109$ and so $(1, 2)$ is a local maximum.

Differentiation

1 Locate and classify the stationary points of the function $f(x) = x^3(x - 2)^2$.

2 Determine the minimum value of the function

$f(x) = \sqrt{x} + \frac{1}{\sqrt{x}}$.

3 Determine the equations of both the tangent and normal to the curve
$y = f(x) = 2x^3 + 3x^2 - 12x + 2$
at the point $(-1, 15)$.

4 Show that the function
$y = f(x) = \ln x - x^2 + x$
attains its maximum value of 0 when $x = 1$.

5 The equation of a curve is given by $y = x^3 + 5x^2 - 4x - 20$.

(a) Find the gradient of the curve at:

(i) the point where it crosses the y-axis

(ii) the points where it crosses the x-axis.

(b) Find and classify the turning points of the curve.

6 Show that the curve $2\sqrt{x}(5 - x)$ has a stationary point at $x = \frac{5}{3}$.

Determine the y-coordinate of the stationary point and whether it is a maximum or minimum.

7 An open cylindrical tank is to hold 5000 litres ($5\,m^3$). What are the dimensions of the tank if its surface area is to be kept to a minimum?

8 For $f(x) = x^3$ use the limit definition $f'(x) = \lim_{h \to 0} \frac{f(x + h) - f(x)}{h}$

to show that the derivative of $f(x)$ is $3x^2$.

The answers are on page 101.

Integration

The area under the graph of a positive function $f(x)$ between $x = a$ and $x = b$ is given by the definite integral $\int_a^b f(x)dx$. Integration is the **inverse** of differentiation. If $f(x) = x^n$ ($n \neq -1$) then:

$\int f(x)dx = \frac{x^{n+1}}{n+1} + c$ where c is the arbitrary **constant of integration**.

This result holds for all rational numbers except $n = -1$. Integrals involving an arbitrary constant are **indefinite integrals**, to distinguish them from **definite integrals** in which the **upper** and **lower limits** of integration are specified.

Polynomials can be integrated term by term since the integral of a sum (or difference) of terms is the sum (or difference) of the integral of each term.

Q A curve has gradient function $\frac{dy}{dx} = 3x^2 - 2$ and passes through the point (0, 2). Find the equation of the curve in the form $y = f(x)$ and determine the area enclosed by this curve, the x-axis and $x = 0$ and $x = 2$. (You may assume $f(x)$ is positive for $x \in [0, 2]$).

$\frac{dy}{dx} = 3x^2 - 2$

Integrating each side gives $y = x^3 - 2x + c$. Since the curve passes through (0, 2) then $2 = 0^3 - 2 \times 0 + c \Rightarrow c = 2$ so the equation is $y = x^3 - 2x + 2$. A sketch of the function shows the area required.

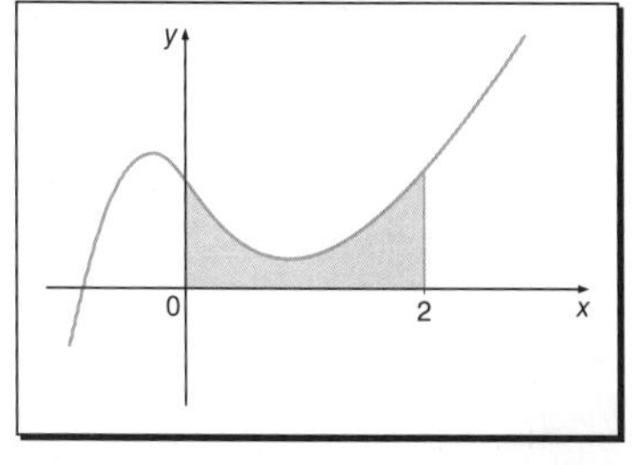

The required area is given by

$$\int_0^2 y dx = \int_0^2 (x^3 - 2x + 2)dx$$

$$= \left[\frac{x^4}{4} - 2\frac{x^2}{2} + 2x\right]_0^2$$

$$= \left(\frac{2^4}{4} - 2 \times \frac{2^2}{2} + 2 \times 2\right) - 0$$

$$= 4$$

You can use integration to find the area enclosed by curves. Take care if the area or any part of it lies below the x-axis. Split the area into regions and if necessary take the modulus of the integral as the area.

Area enclosed by the graph, $x = a$, $x = b$ and the x-axis

$$= \int_a^c f(x)\,dx + \left|\int_c^b f(x)\,dx\right|$$

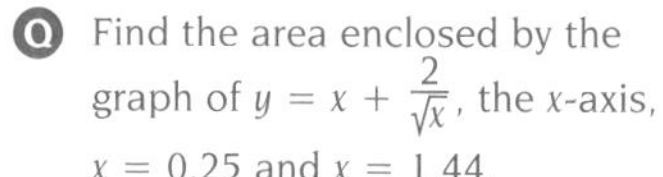

Q Find the area enclosed by the graph of $y = x + \frac{2}{\sqrt{x}}$, the x-axis, $x = 0.25$ and $x = 1.44$.

A sketch of the function shows that it is positive in the region required.

The required area is given by:

$$\int_{0.25}^{1.44}\left(x + \frac{2}{\sqrt{x}}\right)dx = \int_{0.25}^{1.44}(x + 2x^{-\frac{1}{2}})dx$$

$$= \left[\frac{x^2}{2} + \frac{2x^{\frac{1}{2}}}{\frac{1}{2}}\right]_{0.25}^{1.44}$$

$$= \left(\frac{1.44^2}{2} + 4 \times \sqrt{1.44}\right) - \left(\frac{0.25^2}{2} + 4 \times \sqrt{0.25}\right)$$

$$= 3.806$$

Q Find the area enclosed by the function $f(x) = x^3 - 2x^2 - 3x$, the x-axis, $x = -1$ and $x = 3$.

A sketch shows that part of the function is beneath the x-axis so the area needs to be found in two parts, $-1 \leqslant x \leqslant 0$ and $0 \leqslant x \leqslant 3$.

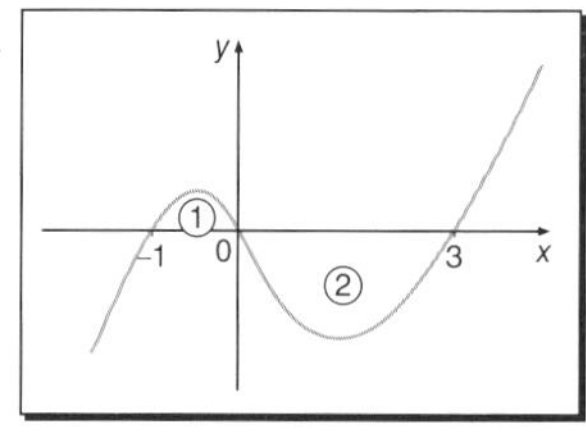

$$\text{Area 1} = \int_{-1}^{0}(x^3 - 2x^2 - 3x)dx = \left[\frac{x^4}{4} - 2\frac{x^3}{3} - 3\frac{x^2}{2}\right]_{-1}^{0} = \frac{7}{12}$$

$$\text{Area 2} = \left|\int_0^3(x^3 - 2x^2 - 3x)dx\right| = \left|\left[\frac{x^4}{4} - 2\frac{x^3}{3} - 3\frac{x^2}{2}\right]_0^3\right| = \left|\frac{-45}{4}\right| = \frac{45}{4}$$

So the total area required $= \frac{7}{12} + \frac{45}{4} = \frac{71}{6}$

Integration of exponential and logarithmic functions

$\int \frac{1}{x}dx = \ln x + c \qquad \int e^{kx}dx = \frac{1}{k}e^{kx} + c$ where k is a constant.

Q Explain why the curve with equation $y = f(x)$, where $f(x) = \frac{1}{x} + e^{2x}$, is positive for $x > 0$. Determine the area of the region enclosed by the curve, the x-axis, $x = 0.2$ and $x = 1$.

Since the exponential function is always positive and since $\frac{1}{x}$ is positive if x is positive then $f(x)$ is positive for $x > 0$.

The area is given by:

$$\int_{0.2}^{1}\left(\frac{1}{x} + e^{2x}\right)dx = \left[\ln x + \tfrac{1}{2}e^{2x}\right]_{0.2}^{1} = (\ln 1 + \tfrac{1}{2}e^{2}) - (\ln 0.2 + \tfrac{1}{2}e^{0.4})$$

$$= 4.56$$

Q Show that the curve $y = \frac{2}{x}$ and the straight line $y = 3 - x$ intersect where $x = 1$ and $x = 2$. Determine the area enclosed between the two curves.

At points where the curves intersect:

$$\frac{2}{x} = 3 - x \Rightarrow 2 = 3x - x^2$$
$$\Rightarrow x^2 - 3x + 2 = 0$$
$$\Rightarrow (x - 1)(x - 2) = 0$$
$$\Rightarrow x = 1, x = 2$$

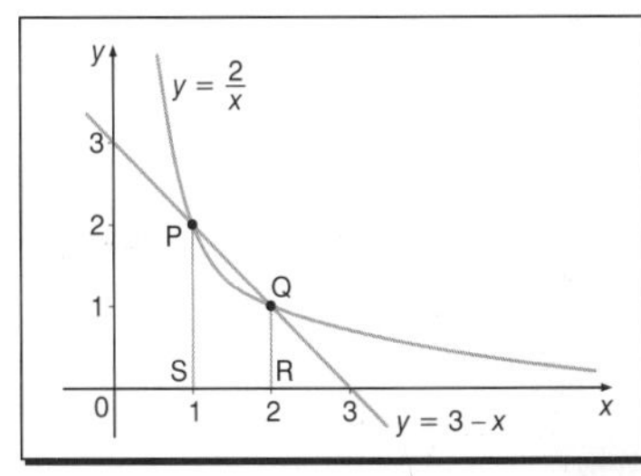

The required area = area of trapezium PQRS – $\int_{1}^{2}\frac{2}{x}dx$

$$= \tfrac{1}{2} \times 1 \times (2 + 1) - \left[2\ln x\right]_{1}^{2}$$
$$= 1.5 - 2\ln 2$$
$$= 0.114$$

Volumes of revolution

The volume generated by rotating the area enclosed by the curve $y = f(x)$, the x-axis, $x = a$ and $x = b$ about the x-axis is given by $V = \pi \int_a^b y^2 dx$.

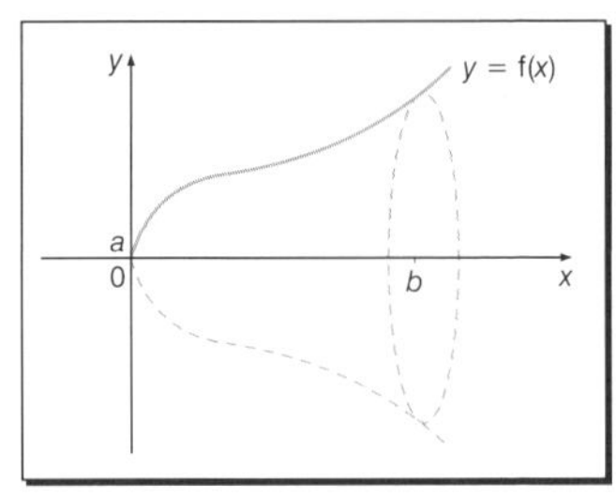

Q Find the volume generated when the area enclosed by the graph of $y = x(4 - x)$ and the x-axis is rotated about the x-axis.

First, make a sketch of the region involved.

The volume of revolution is given by:

$$\pi \int_0^4 y^2 dx = \pi \int_0^4 (x(4 - x))^2 dx$$

$$= \pi \int_0^4 (x^4 - 8x^3 + 16x^2) dx$$

$$= \pi \left[\frac{x^5}{5} - 8\frac{x^4}{4} + 16\frac{x^3}{3} \right]_0^4 = \frac{512\pi}{15}$$

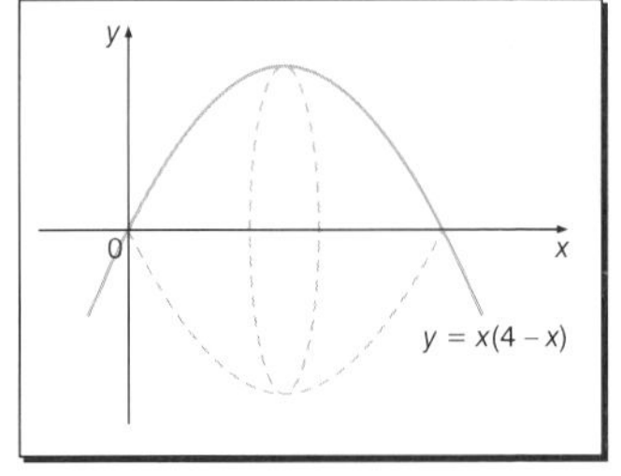

Q Find the volume generated when the area enclosed by the graph of $y = x^2$ is rotated about the y-axis between $y = 0$ and $y = 2$.

First, make a sketch of the region involved.

Note that here the area is being rotated around the y-axis, so the formula needs to be amended accordingly,

$$\text{Volume} = \pi \int_a^b x^2 dy$$

Since $y = x^2$ the volume is given by:

$$\pi \int_0^2 y dy = \pi \left[\frac{y^2}{2} \right]_0^2 = 2\pi$$

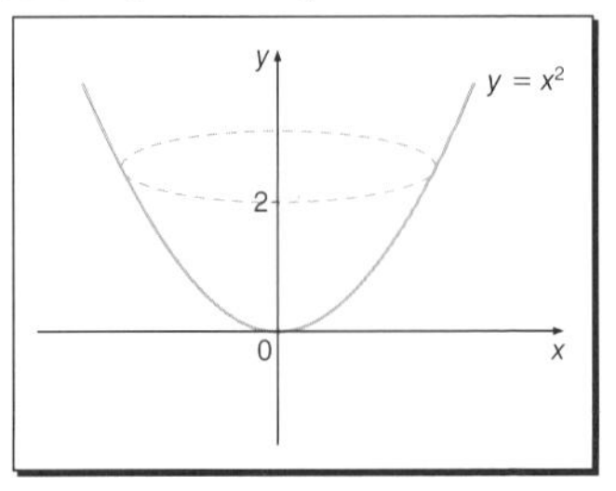

Check yourself

Integration

1 Find the equation of the curve which has the given gradient function and passes through the point shown.

(a) $f'(x) = x^2(x + 1)$ passes through $(1, 1)$

(b) $f'(x) = (x^2 - 1)(3x + 5)$ passes through $(0, -3)$

(c) $f'(x) = 6e^{3x}$ passes through $(0, 3)$

2 Evaluate the following integrals.

(a) $\int_0^4 3\sqrt{x}(1 + x)dx$ **(b)** $\int_2^3 \frac{x^4 - 4}{x^3}dx$

(c) $\int_{-2}^{-1} \frac{(x + 2)^2}{x^4}dx$ **(d)** $\int_1^3 (10x - e^x)dx$

3 Use integration to evaluate the areas of the following regions.

(a) Between $y = 1 - x^2 + x^4$, $y = 0$, $x = 0$ and $x = 1$

(b) Between $y = 2e^{-x}$, the x-axis, $x = -1$ and $x = 1$

4 Find the area enclosed by the curve $y = (x + 2)(x - 1)(x - 2)$, the x-axis, and the lines $x = -2$ and $x = 2$ as shown in the diagram.

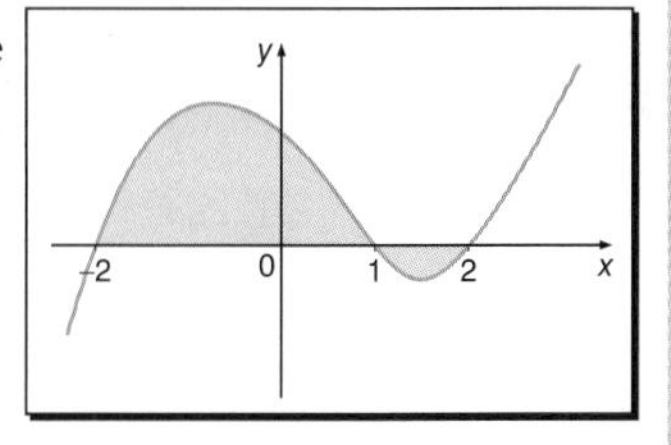

5 Find the volume generated when the area enclosed by the graphs of $y = \frac{3}{x}$ and $y = 4 - x$ is rotated about the y-axis.

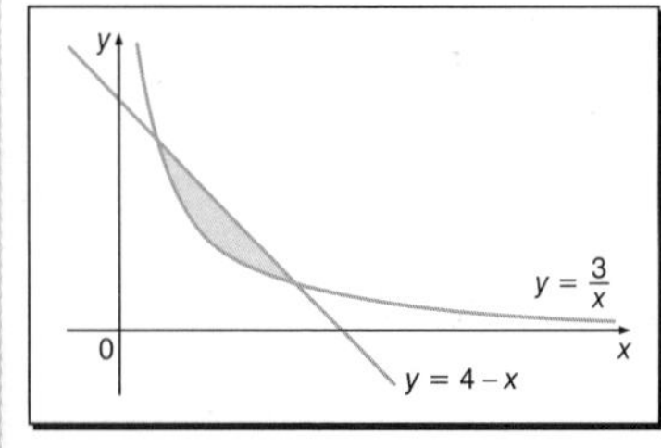

6 Find the volume generated when the area enclosed by the graphs of $y = x$ and $y = x(4 - x)$ is rotated about the x-axis.

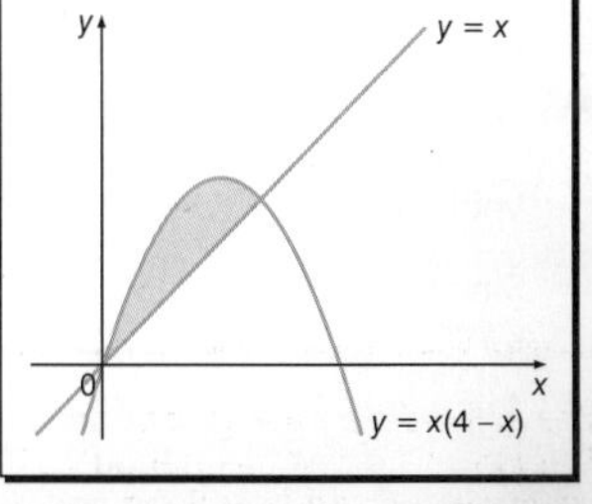

The answers are on pages 102–103.

Sine, cosine and tangent

The **unit circle** is a circle of radius 1, centre the origin, O. The angle θ can be measured in terms of the ratio of two sides of the triangle OPN. The three basic trigonometric functions are **sin**θ, **cos**θ and **tan**θ defined by:

$\sin\theta = \frac{PN}{OP}$, $\cos\theta = \frac{ON}{OP}$ and $\tan\theta = \frac{PN}{ON}$.

The last definition implies $\tan\theta = \frac{\sin\theta}{\cos\theta}$.

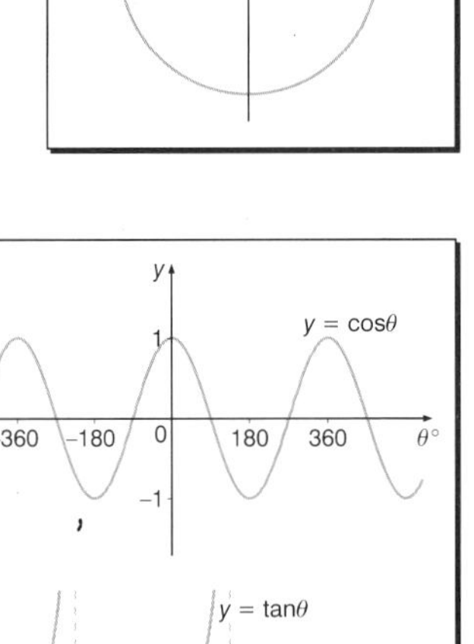

Pythagoras' theorem applied to triangle OPN leads to the trigonometric identity $\sin^2\theta + \cos^2\theta = 1$.

As OP sweeps round the unit circle ($0 \leqslant \theta \leqslant 360°$) so the values of $\sin\theta$ and $\cos\theta$ each span the range $[-1, 1]$ while $\tan\theta$ ranges from $-\infty$ to ∞.
(Note $\tan\theta = 0$ whenever $\sin\theta = 0$.)

The graphs of the functions are like this.

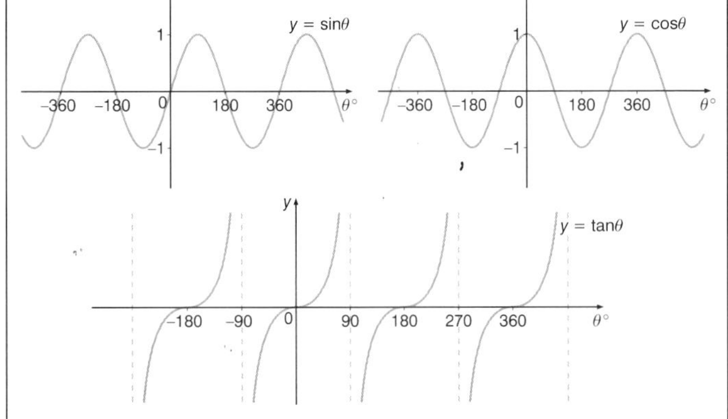

The sine and cosine graphs are wavelike with **amplitude** 1 and **period** 360°. The graph of $\tan\theta$ has period 180°. The sine and tangent functions are examples of **odd functions** and the cosine function is an **even function**.

Radians

As P moves around the circumference of the unit circle it travels a total distance of $2\pi \times 1 = 2\pi$ in each revolution. The angle through which OP turns when P travels 1 unit of length (also equal to the radius) around the circumference of the unit circle is 1 **radian**.

Thus 2π radians $\equiv 360°$ so 1 rad $\equiv \frac{180°}{\pi}$ and $1° \equiv \frac{\pi}{180}$ rad.

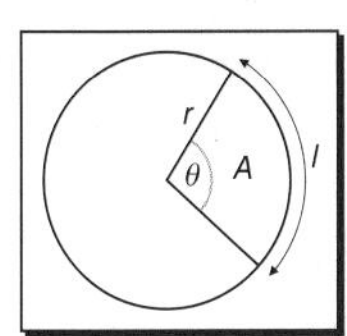

When an angle is expressed without any units it is assumed to be measured in radians.

In a circle of radius r, the **arc length** l subtended at the centre by an angle θ radians is $l = r\theta$. The **area of the corresponding sector** of the circle is $A = \frac{1}{2}r^2\theta$.

Q **(a)** By considering an equilateral triangle of side 2 units show that $\sin 30° = \frac{1}{2}$, $\cos 30° = \frac{\sqrt{3}}{2}$, $\sin 60° = \frac{\sqrt{3}}{2}$ and $\cos 60° = \frac{1}{2}$.

(b) By considering a right-angled isosceles triangle, for which the two equal sides have length 1 unit, show that $\sin 45° = \cos 45° = \frac{1}{\sqrt{2}}$.

(a) N is the foot of the perpendicular from A to BC.

$\sin 30° = \frac{BN}{AB} = \frac{1}{2}$ and $\cos 60° = \frac{BN}{AB} = \frac{1}{2}$

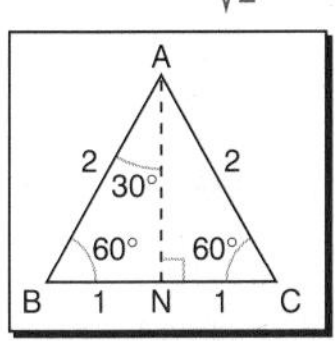

Using Pythagoras' theorem in triangle ABN gives:
$AN = \sqrt{2^2 - 1^2} = \sqrt{3}$

$\cos 30° = \frac{AN}{AB} = \frac{\sqrt{3}}{2}$ and $\sin 60° = \frac{AN}{AB} = \frac{\sqrt{3}}{2}$

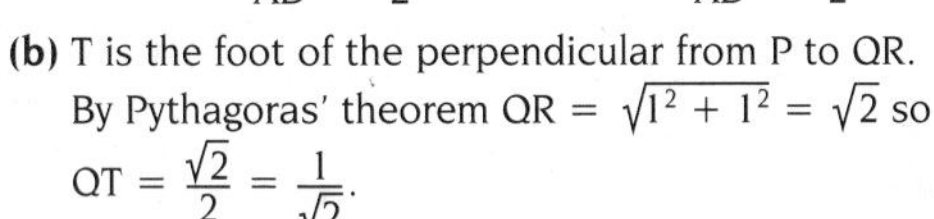

(b) T is the foot of the perpendicular from P to QR. By Pythagoras' theorem $QR = \sqrt{1^2 + 1^2} = \sqrt{2}$ so $QT = \frac{\sqrt{2}}{2} = \frac{1}{\sqrt{2}}$.

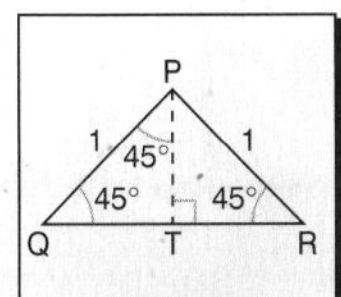

So $\sin QPT = \sin 45° = \frac{QT}{PQ} = \frac{1}{\sqrt{2}}$

and $\cos PQT = \cos 45° = \frac{QT}{PQ} = \frac{1}{\sqrt{2}}$

These results illustrate the fact that
$\cos(90° - \theta) = \sin\theta$ and $\sin(90° - \theta) = \cos\theta$.

Q A cylindrical pipe of radius 10 cm contains water to a depth of 16 cm. Find the cross-sectional area of water.

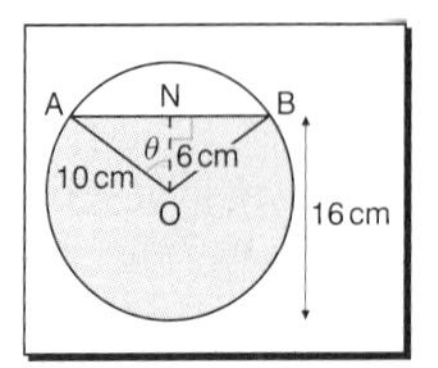

Area of complete circle $= \pi \times 10^2 = 100\pi \text{ cm}^2$

Area of shaded region = area of complete circle – area of sector OAB + area of triangle OAB

From the diagram, $\cos\theta = \frac{6}{10} \Rightarrow \theta = 0.927$ rad.

Thus area of sector $= \frac{1}{2} \times 10^2 \times (2 \times 0.927) = 92.7 \text{ cm}^2$.

Using Pythagoras' theorem in triangle OAN:

$AN = \sqrt{10^2 - 6^2} = 8$ so the area of triangle OAB $= 48 \text{ cm}^2$.

The cross-sectional area of water $= 100\pi - (92.7 - 48) = 269.5 \text{ cm}^2$.

Trigonometric equations

A **trigonometric equation** is an equation that involves any or all of the sine, cosine and tangent functions. To solve such an equation you need to find all the values of the angle that satisfy the equation. This will usually involve the **inverse trigonometric functions** $\sin^{-1}$, $\cos^{-1}$, $\tan^{-1}$. Your calculator displays the principal value, in the range $-90° \leqslant \theta \leqslant 180°$ (or $-\frac{\pi}{2} \leqslant \theta \leqslant \pi$). e.g. $\cos^{-1}0.5 = 60°$, $\tan^{-1}1 = 45°$, $\sin^{-1}-\frac{\sqrt{3}}{2} = -60°$.

Q Find values of θ, $0 \leqslant \theta \leqslant 360°$, such that $2 - 10\sin\theta = 7$.

$2 - 10\sin\theta = 7 \Rightarrow \sin\theta = -\frac{1}{2} \Rightarrow \theta = -30°$ (from the calculator) which is not in the required range.

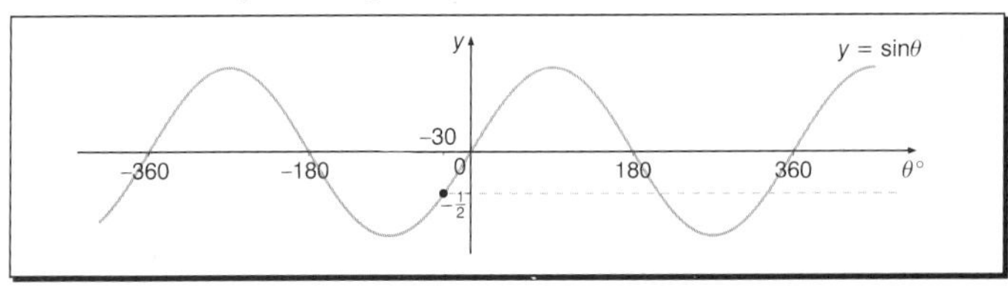

The graph of $\sin\theta$ shows that $\sin\theta$ is $-\frac{1}{2}$ for two values in the required range.

The two values are $\theta = 180° + 30° = 210°$ and $360° - 30° = 330°$.

Q Show that the equation $15\sin^2\theta = 13 + \cos\theta$ may be written as a quadratic equation in $\cos\theta$. Hence solve the equation for θ in the range $[0, 360°]$.

Since $\sin^2\theta + \cos^2\theta = 1$ then $\sin^2\theta = 1 - \cos^2\theta$.

The equation becomes:

$15(1 - \cos^2\theta) = 13 + \cos\theta \Rightarrow 15\cos^2\theta + \cos\theta - 2 = 0$

Letting $x = \cos\theta$ gives a quadratic:

$15x^2 + x - 2 = 0 \Rightarrow (3x - 1)(5x + 2) = 0 \Rightarrow x = \cos\theta = \frac{1}{3}$ or $-\frac{2}{5}$

$\theta = \cos^{-1}\frac{1}{3} = 70.5°$ (from calculator) and $360° - 70.5° = 289.5°$

$\theta = \cos^{-1}-\frac{2}{5} = 113.6°$ (from calculator) and $180° + 66.4° = 246.4°$

Thus $\theta = 70.5°, 113.6°, 246.4°, 289.5°$

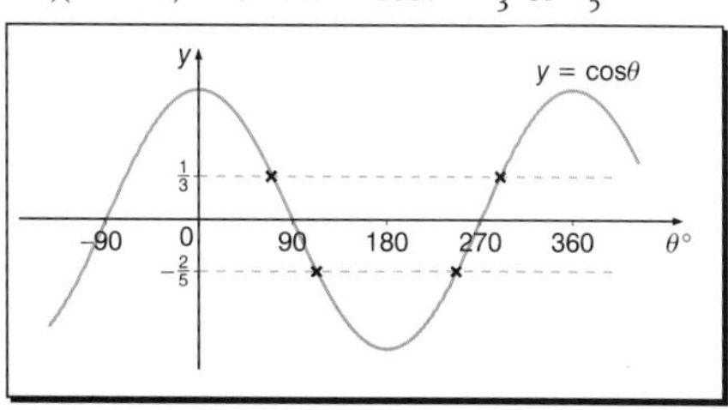

Compound angle formulae

$\sin(\theta \pm \phi) = \sin\theta\cos\phi \pm \sin\phi\cos\theta$ $\quad$ $\cos(\theta \pm \phi) = \cos\theta\cos\phi \mp \sin\theta\sin\phi$

$$\tan(\theta \pm \phi) = \frac{\tan\theta \pm \tan\phi}{1 \mp \tan\theta\tan\phi}$$

Double angle formulae

You can obtain these by setting $\phi = \theta$ in the compound angle formulae.

$\sin 2\theta = 2\sin\theta\cos\theta$ $\quad$ $\cos 2\theta = \cos^2\theta - \sin^2\theta = 2\cos^2\theta - 1 = 1 - 2\sin^2\theta$

$$\tan 2\theta = \frac{2\tan\theta}{1 - \tan^2\theta}$$

Q Use the compound angle formulae (not a calculator) to find:
(a) $\sin 15°$ **(b)** $\tan 105°$ using the values you found in the example on page 40, and leaving your answer in surds.

(a) $15° = 45° - 30°$

$\sin 15° = \sin(45° - 30°) = \sin 45°\cos 30° - \sin 30°\cos 45°$

$$= \frac{1}{\sqrt{2}} \times \frac{\sqrt{3}}{2} - \frac{1}{2} \times \frac{1}{\sqrt{2}} = \frac{1}{2\sqrt{2}}(\sqrt{3} - 1)$$

(b) $\tan 105° = \tan(45° + 60°) = \dfrac{\tan 45° + \tan 60°}{1 - \tan 45°\tan 60°} = \dfrac{1 + \sqrt{3}}{1 - \sqrt{3}}$

$= -(2 + \sqrt{3})$

Q Solve the equation $3\cos 2x = \cos x - 2$ in the range $0 \leqslant x \leqslant 360°$.

Using the double angle formula to replace $\cos 2x$:

$6\cos^2 x - \cos x - 1 = 0 \Rightarrow (3\cos x + 1)(2\cos x - 1) = 0 \Rightarrow \cos x = -\frac{1}{3}$ or $\frac{1}{2}$.

$\cos x = -\frac{1}{3} \Rightarrow x = 109.5°$ (from calculator) or $x = 360° - 109.5° = 250.5°$

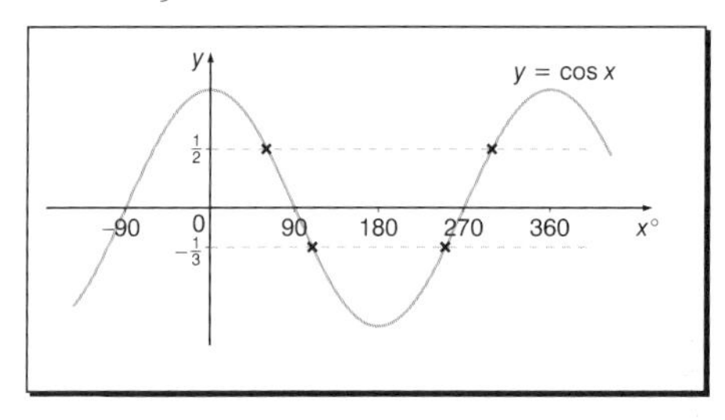

$\cos x = \frac{1}{2} \Rightarrow x = 60°$ or $300°$.

So $x = 60°, 109.5°, 250.5°, 300°$.

Combining trigonometric functions

The compound angle formulae can be used to express $a\sin x + b\cos x$ as a single trigonometric term as either:

$a\sin x + b\cos x = R\cos(x \pm \alpha)$ or

$a\sin x + b\cos x = R\sin(x \pm \alpha)$

Let $3\sin x - 4\cos x = R\sin(x + \alpha) = R\sin x\cos\alpha + R\cos x\sin\alpha$

If $x = 0$: $R\sin\alpha = -4$

If $x = 90°$: $R\cos\alpha = 3$

$(R\sin\alpha)^2 + (R\cos\alpha)^2 = R^2 = (-4)^2 + 3^2 \Rightarrow R = 5$

$\frac{R\sin\alpha}{R\cos\alpha} = \tan\alpha = -\frac{4}{3}$

From above, $\sin\alpha$ is negative and $\cos\alpha$ is positive, so α is in the fourth quadrant. Therefore $\alpha = 360° - 53.1° = 306.9°$

So $3\sin x - 4\cos x = 5\sin(x + 306.9°)$

Reciprocal trigonometric functions

The reciprocal trigonometric functions are:

$\text{cosec}x = \dfrac{1}{\sin x}$ $\sec x = \dfrac{1}{\cos x}$ $\cot x = \dfrac{1}{\tan x}$

The trigonometric identity $\sin^2 x + \cos^2 x = 1$ has already been established. Dividing this throughout by $\sin^2 x$ leads to the identity $\text{cosec}^2 x = 1 + \cot^2 x$.

Dividing by $\cos^2 x$ leads to the identity $\sec^2 x = 1 + \tan^2 x$.

You can use a graphics calculator to display the graphs of $y = \sin x$ and $y = \text{cosec}x$ on the same axes.The **asymptotes** of cosec$x$ correspond to the zeros of sin$x$. The two curves touch repeatedly, when $x = \frac{\pi}{2}, \frac{3\pi}{2}, \frac{5\pi}{2}, \ldots$ since at these points $\sin x = \pm 1$, hence $\text{cosec}x = \pm 1$.

Similar behaviour is exhibited by $y = \cos x$ and $y = \sec x$.

You can also plot the graphs of $y = \tan x$ and $y = \cot x$ on a graphics calculator. Note that the asymptotes of one are the zeros of the other.

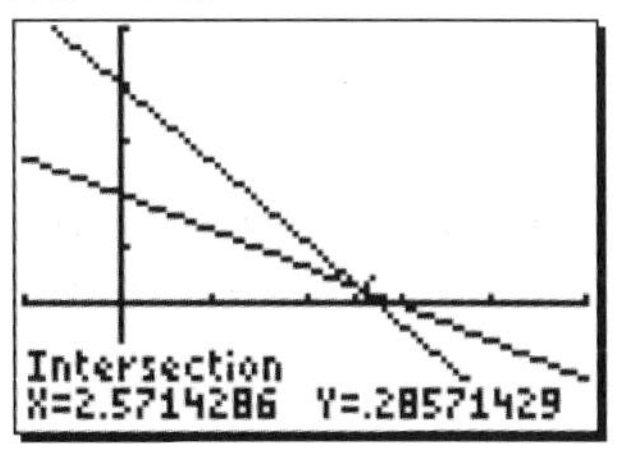

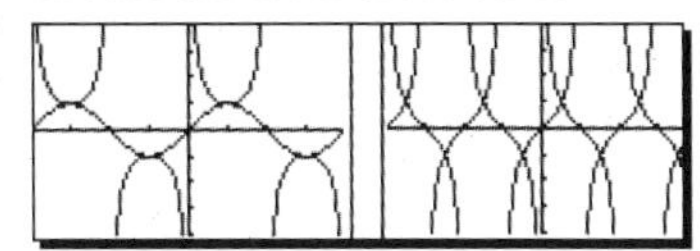

Q Find values of θ for which $\text{cosec}2\theta = 2$, in the range $0 \leqslant \theta \leqslant 360°$.

$\text{cosec}2\theta = 2 \Rightarrow \dfrac{1}{\sin 2\theta} = 2 \Rightarrow \sin 2\theta = 0.5 \Rightarrow 2\theta = \sin^{-1} 0.5 = 30°$

$0 \leqslant \theta \leqslant 360° \Rightarrow 0 \leqslant 2\theta \leqslant 720° \Rightarrow 2\theta = 30°, 180° - 30°, 360° + 30°, 360° + 150°$

So $\theta = 15°, 75°, 195°, 255°$.

Q Prove that $\dfrac{\text{cosec}\theta}{\text{cosec}\theta - \sin\theta} = \sec^2\theta$.

$$\text{Left-hand side} = \frac{\frac{1}{\sin\theta}}{\frac{1}{\sin\theta} - \sin\theta} = \frac{\frac{1}{\sin\theta}}{\frac{1 - \sin^2\theta}{\sin\theta}} = \frac{\frac{1}{\sin\theta}}{\frac{\cos^2\theta}{\sin\theta}}$$

which simplifies to $\dfrac{1}{\cos^2\theta} = \sec^2\theta$ which is equal to the right-hand side.

Check yourself

Trigonometry

1 Show that the equation $2\cos^2 x + 3\sin x = 0$ can be written as a quadratic in $\sin x$. Hence solve the equation for x in the range $0 \leqslant x \leqslant 360°$.

2 If $F(\cos\alpha + \mu\sin\alpha) = \mu W$ where $\mu = \tan\lambda$ prove that $F = \dfrac{W\sin\lambda}{\cos(\alpha - \lambda)}$.

3 Use compound angle formulae to show that $\sin 3\theta = 3\sin\theta - 4\sin^3\theta$ and hence solve the equation $\sin 3\theta - 2\sin\theta = 0$ for θ in the range $0 \leqslant \theta \leqslant 360°$.

4 Prove that $\dfrac{2\sin\theta + \sin 2\theta}{1 - \cos 2\theta} = \dfrac{\sin\theta}{1 - \cos\theta}$.

5 Simplify the expression $\sqrt[3]{\dfrac{\sin\theta}{1 + \cot^2\theta}}$.

6 An angle θ subtends an arc AB of length 12 cm on a circle of radius r cm. The area of the sector OAB (O is the centre of the circle) is 54 cm^2. Form two equations in r and θ and hence find their values.

7 The height h m of the tide at time t hours after midnight on a certain day is given by $h = 5 + 3\sin\dfrac{\pi t}{6}$.
Sketch a graph of h (vertically) against t over a day.
At what time do the high and low tides occur?
When is the tide first at a height of 6 m and rising?

8 A wire of length 50 cm is bent into the shape of a sector of a circle of radius r cm and angle θ radians. Find a formula for the area of the sector. Find the value of r and the corresponding value of θ which gives the largest sector area.

9 Find a single function equivalent to $f(x) \equiv 5\cos x - 12\sin x$. Determine the maximum and minimum values of $f(x)$.

10 Find values of θ in the range $0 \leqslant \theta \leqslant 360°$ for which (a) $\operatorname{cosec}\theta = 4$ (b) $\sec 2\theta = 5$.

The answers are on pages 104–105.

Sequences and series

A **sequence** is a list of numbers such as 1, 1, 2, 3, 5, 8, . . . When the terms of a sequence are added together, the result is a **series**. A general sequence can be represented as $u_1, u_2, u_3, \ldots$ where u_k denotes the kth term.

S_n denotes the sum of the first n terms of the series $u_1 + u_2 + u_3 + \ldots + u_n$.

$S_n = \sum_{k=1}^{n} u_k$

Q **(a)** A sequence is defined inductively as $u_{k+1} = u_k + 2$; $u_1 = 1$. Write out the first five terms of the sequence and determine S_5.

(b) Write the first four terms of the sequence defined by $u_k = \frac{(-1)^k}{\sqrt{k}}$. What happens to u_k as $k \to \infty$?

(a) $u_1 = 1$, $u_2 = u_1 + 2 = 1 + 2 = 3$, $u_3 = u_2 + 2 = 3 + 2 = 5$, $u_4 = 5 + 2 = 7$, $u_5 = u_4 + 2 = 9$

Then $S_5 = 1 + 3 + 5 + 7 + 9 = 25$

(b) $u_1 = \frac{(-1)^1}{\sqrt{1}} = -1$, $u_2 = \frac{(-1)^2}{\sqrt{2}} = \frac{1}{\sqrt{2}}$, $u_3 = \frac{(-1)^3}{\sqrt{3}} = \frac{-1}{\sqrt{3}}$, $u_4 = \frac{(-1)^4}{\sqrt{4}} = \frac{1}{2}$,

As k gets larger the terms get smaller so $u_k \to 0$ as $k \to \infty$.

Arithmetic sequences and series (or progressions)

In an **arithmetic sequence** the difference between successive terms is constant. If the first term is denoted by a and the common difference is d then the sequence is $a, a + d, a + 2d, \ldots$

Defined **inductively** the sequence is $u_{k+1} = u_k + d$; $u_1 = a$ and as a formula $u_k = a + (k - 1)d$. The sum of the first n terms is:

$S_n = \frac{n}{2}\left[2a + (n - 1)d\right] = \frac{1}{2}(a + l)d$ where l is the last term.

Q The third term of an AP (arithmetic progression, or series) is 11 and the eighth term is 31. Determine the first term, the common difference and the sum of the first 10 terms.

$u_3 = 11 = a + 2d$ **(i)** and $u_8 = 31 = a + 7d$ **(ii)**. Solving equations **(i)** and **(ii)** for a and d: **(ii)** – **(i)** $\Rightarrow 5d = 20 \Rightarrow d = 4$.

Substitute into **(i)** $\Rightarrow a = 11 - 2 \times 4 = 3$.

The first term is 3, the common difference is 4.

$S_{10} = \frac{10}{2}\left[2 \times 3 + (10 - 1) \times 4\right] = 210$

Geometric sequences and series (or progressions)

In a **geometric sequence** the ratio of each term to the preceding term is constant. If the first term is a and the common ratio is r then the sequence is $a, ar, ar^2, ar^3, \ldots$. The geometric series is defined inductively as $u_{k+1} = ru_k$, $u_1 = a$ and as a formula $u_k = ar^{k-1}$. The sum of the first n terms is $S_n = \frac{a(1-r^n)}{1-r}$. If $|r| < 1$ then $r^n \to 0$ as $n \to \infty$ and the sum to infinity of the series is $S_\infty = \frac{a}{1-r}$.

Q The second term of a GP (geometric progression, or series) is 2500 and the fifth term is 20. Find the first term, the common ratio and the sum to infinity.

$u_2 = 2500 = ar^1$ **(i)**, $u_5 = 20 = ar^4$ **(ii)**.

Dividing **(ii)** by **(i)**: $r^3 = \frac{20}{2500} = \frac{1}{125} \Rightarrow r = \frac{1}{5}$.

Substituting for r into **(i)**: $2500 = \text{a} \times \frac{1}{5} \Rightarrow a = 12\,500$.

The first term of the series is 12 500 and the common ratio is $\frac{1}{5}$.

Since the common ratio is less than 1 then the sum to infinity exists and is $S_\infty = \frac{12\,500}{1 - \frac{1}{5}} = 15\,625$.

Binomial theorem

The expansion of $(a + b)^n$ when n is a positive integer is:

$$(a+b)^n = a^n + {}^nC_1 a^{n-1}b + {}^nC_2 a^{n-2}b^2 + \ldots + {}^nC_k a^{n-k}b^k + \ldots + {}^nC_{n-1} ab^{n-1} + b^n$$

where ${}^nC_k = \frac{n!}{k!(n-k)!}$ denotes the **binomial coefficient**.

For a given value of n you can find the binomial coefficients from the $(n - 1)$th row of Pascal's triangle.

Binomial series

This is a special case of the binomial theorem when $a = 1$ and $b = x$. By evaluating the binomial coefficients ${}^nC_k$ you can obtain:

$$(1+x)^n = 1 + nx + \frac{n(n-1)}{2!}x^2 + \frac{n(n-1)(n-2)}{3!}x^3 + \ldots + nx^{n-1} + x^n$$

If n is fractional or negative the binomial series becomes an infinite series (it has an infinite number of terms) and is convergent provided $|x| \leqslant 1$.

Q For the expression $(a + b)^{10}$:

(a) find the coefficient of a^4b^6

(b) write down the first four terms in descending powers of a.

(a) From the definition, the coefficient of a^4b^6 is:

$${}^{10}C_6 = \frac{10!}{6!4!} = \frac{10 \times 9 \times 8 \times 7}{4 \times 3 \times 2 \times 1} = 210$$

(b) The first four terms are:

$$a^{10} + {}^{10}C_1a^9b + {}^{10}C_2a^8b^2 + {}^{10}C_3a^7b^3$$
$$= a^{10} + 10a^9b + 45a^8b^2 + 120a^7b^3$$

Q Expand the following expressions in ascending powers of x up to the term in x^4.

(a) $(1 + 3x)^8$ **(b)** $(1 - x)^6$ **(c)** $(1 + x^2)^4$

(a) Let $z = 3x$ so the expression becomes $(1 + z)^8$.
Apply the binomial series to obtain:

$$(1 + z)^8 = 1 + 8z + 28z^2 + 56z^3 + 70z^4 + \ldots$$

Finally replace z with $3x$ to give:

$$(1 + z)^8 = 1 + 8(3x) + 28(3x)^2 + 56(3x)^3 + 70(3x)^4 + \ldots$$
$$= 1 + 24x + 252x^2 + 1512x^3 + 5670x^4 \ldots$$

(b) $(1 - x)^6 = (1 + (-x))^6 = 1 + 6(-x) + 15(-x)^2 + 20(-x)^3 + 15(-x)^4 \ldots$
$= 1 - 6x + 15x^2 - 20x^3 + 15x^4 + \ldots$

(c) $(1 + x^2)^4 = (1 + (x^2))^4 = 1 + 4(x^2) + 6(x^2)^2 + \ldots$
$= 1 + 4x^2 + 6x^4 + \ldots$

Q Explain why, if h is sufficiently small, $(1 + h)^n \approx 1 + nh$. The equation $x^3 - 1.05x^2 - 4x + 4.2 = 0$ is known to have a solution near $x = 1$. Write the solution in the form $x = 1 + h$ and substitute in the equation, then use the above approximation to obtain a better estimate of the solution.

Using the binomial series $(1 + h)^n = 1 + nh + \ldots$ so if h very small (so that h^2 and higher powers can be neglected) then $(1 + h)^n \approx 1 + nh$.
Letting $x = 1 + h$, the cubic becomes

$$(1 + h)^3 - 1.05(1 + h)^2 - 4(1 + h) + 4.2 = 0$$

$$1 + 3h - 1.05(1 + 2h) - 4(1 + h) + 4.2 = 0 \Rightarrow h = \tfrac{0.15}{3.1} = 0.0484$$

A new estimate for the root is thus $x = 1 + h = 1.0484 = 1.05$ (3 s.f.).

Check yourself

Sequences and Series

1 The sum of the first n terms of an AP is given by $S_n = 2n^2 + 3n$.

(a) Show that the nth term of the series, $u_n = S_n - S_{n+1}$, and hence show that $u_n = 4n + 1$.

(b) If the first term of the series is a and the common difference is d, find the values of a and d.

2 The second term of a GP is 40 and the fifth term is 1.08.

(a) Show that the common ratio of the series is 0.3.

(b) Determine the first term of the series.

(c) Calculate the sum to infinity of the series, giving your result as an exact fraction.

(d) Find the value of n for which S_n, the sum of the first n terms of the GP, satisfies the inequality $|S_n - S_\infty| < 0.001$.

3 Find the first four terms in the expansion of $(1 + 3x)^{-2}$. State the range of values of x for which the expansion converges. Use your result to estimate $\frac{1}{1.03^2}$ to 3 s.f.

4 Show that if x is so small in comparison to 1 that x^3 and higher powers can be ignored then:

$$\frac{(1 - 4x)^{\frac{1}{2}}(1 + 3x)^{\frac{1}{3}}}{(1 + x)^{\frac{1}{2}}} = 1 - \tfrac{3}{2}x - \tfrac{33}{8}x^2$$

5 Identify whether each of the following sequences is arithmetic or geometric. For each of them, find the 20th term and the sum of the first 20 terms.

(a) 100, 90, 81, 72.9, 65.61, ...

(b) 5, 12, 19, 26, 33, ...

6 Find the first four terms of the expansion of $(1 + 2x)^{-2}$ and state the range of values of x for which it converges.

The answers are on pages 106–107.

Domain and range

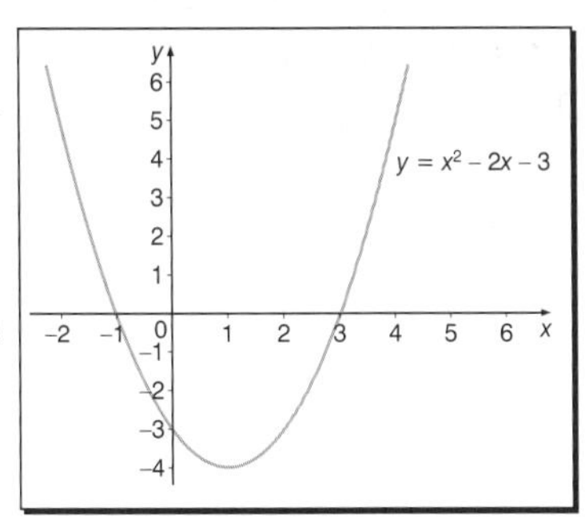

A **function** is a 1–1 **mapping** from a set of values called the **domain** to a set of values called the **range** (or **codomain**).

The diagram shows the graph of $y = x^2 - 2x - 3$.

For the domain $x \in [0, 3]$ the range is $[-4, 0]$ but for the domain $x \in [-1, 0]$ the range is $[-3, 0]$.

For $x \in R$ the range is $[-4, \infty]$.

The notation f(x) is used to indicate that the function with the name f is dependent on the variable x. The above quadratic could thus be written as $f(x) = x^2 - 2x - 3$ or alternatively $f{:}x \rightarrow x^2 - 2x - 3$.

Composite functions

Functions can be combined to produce **composite** functions. Usually the order in which the functions are combined is important. For example if $f(x) = x^2$ and $g(x) = \sin x$, each with the domain R, then
$f(g(x)) = f(\sin x) = (\sin x)^2 = \sin^2 x$ whereas
$g(f(x)) = g(x^2) = \sin(x^2)$ and clearly $fg \neq gf$.

For the composite function gf, apply f first, then apply g to this result.

Q Use a graphics calculator to obtain the graph of:
(a) $f(x) = 18x - 2x^2, x \in R$
(b) $f(x) = |2 - 3x|, x \in [-2, 2]$.

What is the range in each case?

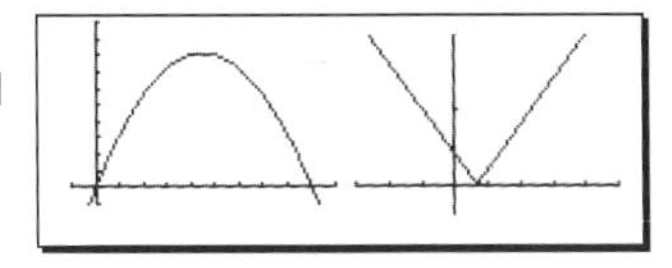

(a) $y = f(x), x \in R$, range $(-\infty, 40.5]$
(b) $y = f(x)$, range $[0, 4)$

Q Use a graphics calculator to obtain the graph of $y = f(x) = 2 + x - x^2$. What is the range corresponding to:
(a) the domain $[0, 3]$
(b) the domain $[-1, 2]$?

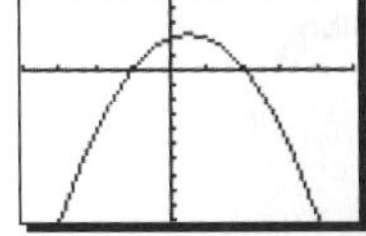

(a) The range is $[-4, 2.25]$.
(b) The range is $[0, 2.25]$.

Inverse functions

The inverse of a function f(x) is the function, denoted by $f^{-1}(x)$ which when applied to f(x) gives the original x-value.

$$x \underset{f^{-1}}{\overset{f}{\rightleftarrows}} f(x)$$

Don't confuse the inverse $f^{-1}(x)$ with the reciprocal of f(x) i.e. $\frac{1}{f(x)}$.

You can think of the inverse function $f^{-1}(x)$ as undoing the effect of the function f(x). For example if f(x) is the squaring function $f(x) = x^2$ then its inverse $f^{-1}(x)$ is the square root function.

Q Find the inverse of the function $f(x) = \frac{2}{x-1}$, $x \in R$, $x \geqslant 1$.

Let $y = f(x) = \frac{2}{x-1}$, then $xy - y = 2 \Rightarrow xy = 2 + y \Rightarrow x = \frac{2}{y} + 1$.

So the inverse function is $f^{-1}(x) = 1 + \frac{2}{x}$.

If you draw the graphs of $y = f(x)$, $y = f^{-1}(x)$ and $y = x$ on the same axes, the function and its inverse are symmetric with respect to the line $y = x$.

Q For the function in the example above, confirm that $ff^{-1}(x) = x$.

The composite notation $ff^{-1}(x)$ means $f(f^{-1}(x))$.

Now $f(f^{-1}(x)) = \frac{2}{f^{-1}(x) - 1} = \frac{2}{1 + \frac{2}{x} - 1} = x$

(It is also the case that $f^{-1}f(x) = x$).

Transforming functions

The transformation $y = af(x + b) + c$ is a **translation** of the function $y = f(x)$ through $-b$ units parallel to the x-axis followed by a **stretch**, factor a, in the y-direction followed by a **translation** of $+c$ units parallel to the y-axis.

Q Show how the graph of the curve $y = f(x) = 2x^2 + 4x - 1$ can be obtained from the graph of $y = x^2$ using translations and stretches.

First complete the square on $f(x)$.

$f(x) = 2(x^2 + 2x) - 1 = 2((x + 1)^2 - 1) - 1 = 2(x + 1)^2 - 3$

So the basic quadratic $y = x^2$ is first shifted through 1 unit to the left parallel to the x-axis, then stretched by a factor 2 in the y-direction (which has the effect of steepening the quadratic curve) and finally shifted 3 units downwards and parallel to the y-axis.

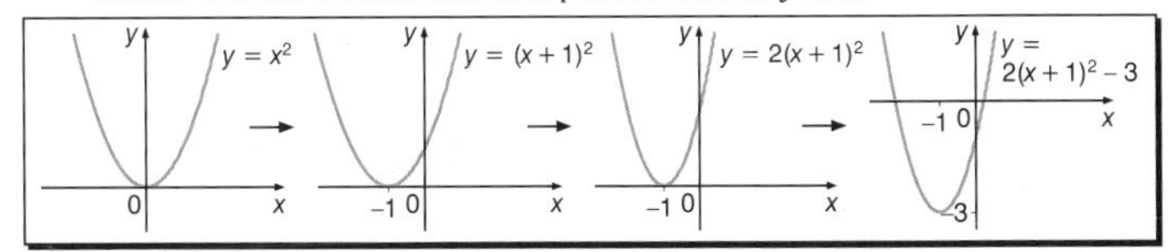

Transforming trigonometric functions

The same ideas can be applied to trigonometric functions. For example, the function $y = a\sin(b(x + c)) + d$ represents an oscillation of amplitude a which has been shifted $+d$ units parallel to the y-axis. Relative to the basic sine wave $y = \sin x$ (of period 2π) the above curve has been translated $-c$ units parallel to the x-axis and has also experienced a stretch, factor b, in the x-direction which has the effect of changing the period to $\frac{2\pi}{b}$.

Q Sketch the graphs of $y = \cos x$ and $y = 0.5\cos(2(x + \frac{\pi}{4})) - 1$ for $0 \leqslant x \leqslant 2\pi$.

The modified cosine function has period $\frac{2\pi}{2}$ so it is compressed (a fractional stretch) in the x-direction so that two cycles occur on $[0, 2\pi]$. The amplitude is 0.5, the complete waveform is shifted $-\frac{\pi}{4}$ units parallel to the x-axis and 1 unit downwards, parallel to the y-axis.

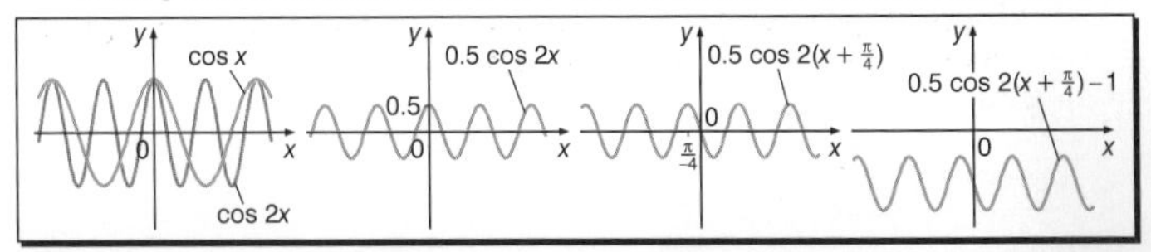

Even and odd functions

A function f(x) is **even** if f($-x$) = f(x) for all $x \in$ domain of f.

A function f(x) is **odd** if f($-x$) = $-$f(x) for all $x \in$ domain of f.

Functions do not have to be either even or odd. For example, consider the function $g(x) = x^2 + x^3$, then $g(-x) = (-x)^2 + (-x)^3 = x^2 - x^3$ and this is not equal to g(x) or $-$g(x). Hence g(x) is neither odd nor even.

Even functions ($y = x^2$, $y = \cos x$, ...) have the y-axis as a line of symmetry.

Odd functions ($y = x^3$, $y = \sin x$, ...) have half-turn rotational symmetry about (0, 0).

Q Say whether each of the following functions is even, odd or neither.

(a) $y = x^3 + 1$ **(b)** $y = \frac{2}{x^3} + \frac{1}{x}$ **(c)** $y = x^6 - x^4$

(a) Let $f(x) = x^3 + 1$ then $f(-x) = (-x)^3 + 1 = -x^3 + 1$.
So $f(-x) \neq f(x)$ and $f(-x) \neq -f(x)$, f(x) is neither even nor odd.

(b) $f(x) = \frac{2}{x^3} + \frac{1}{x}$ so $f(-x) = \frac{2}{-x^3} + \frac{1}{-x} = -\left(\frac{2}{x^3} + \frac{1}{x}\right) = -f(x)$
So f(x) is odd.

(c) $f(x) = x^6 - x^4$ so $f(-x) = (-x)^6 - (-x)^4 = x^6 - x^4 = f(x)$
So f(x) is even.

Q The function f(x) is an even function defined on the interval [-3, 3]. Given that $f(x) = 1$, $0 \leqslant x \leqslant 1$ and $f(x) = 2x - 1$, $1 \leqslant x \leqslant 3$:
(a) sketch the graph of f(x) for $-3 \leqslant x \leqslant 3$
(b) find the values of x for which $f(x) = 2$.

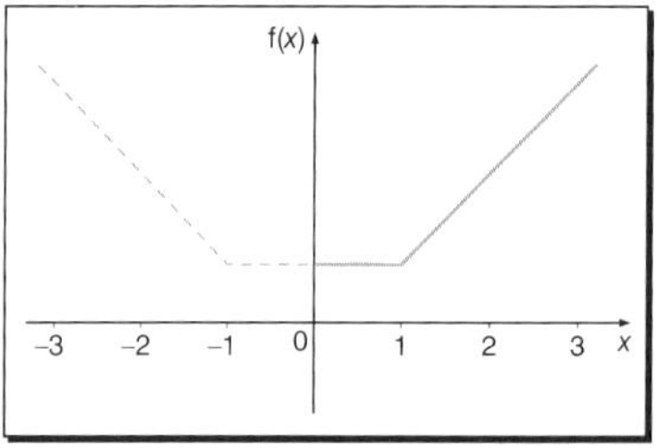

(a) Since f is even, its graph will be symmetric with respect to the y-axis so the graph for $-3 \leqslant x \leqslant 0$ is the reflection of the solid line in the vertical axis.

(b) When $f(x) = 2$ there are two possible (equal and opposite) values of x. Either read the x-coordinates off the graph as $x = \pm 1.5$ or alternatively solve $2 = 2x - 1 \Rightarrow x = 1.5$ and then use symmetry to deduce the other solution $x = -1.5$.

Check yourself

Functions

1 The functions f and g are defined by $f:x \to 3x + 1, x \in R$ and $g:x \to x^2 - 2, x \in R$.

(a) Find the function fg.

(b) What is the range of fg?

(c) Solve the equation $f^{-1}(x) = g(x)$.

2 The functions f and g are defined by $f(x) = \frac{x^3}{3}$ and $g(x) = x - 2$.

(a) Find the function fg.

(b) Find f^{-1} and g^{-1}.

(c) Confirm that the inverse of fg is related to the inverses of f and g by the equation $(fg)^{-1} = g^{-1}f^{-1}$.

3 Confirm that the function $y = f(x) = \frac{3}{x^2}$ is an even function and sketch its graph. For each of the following transformations sketch the transformed curve, give its equation and state the equations of the asymptotes.

(a) translation of -1 unit parallel to the x-axis

(b) translation of $+2$ units parallel to the y-axis

(c) stretch, factor 2, parallel to the y-axis

(d) reflection in the x-axis followed by a translation of $+2$ units parallel to the x-axis.

4 The function f is an odd function defined on the interval $[-2\pi, 2\pi]$. If $f(x) = \pi - x$, $0 < x \leqslant \pi$ and $f(x) = -\sin x$, $\pi \leqslant x \leqslant 2\pi$:

(a) sketch the graph of f for $-2\pi \leqslant x \leqslant 2\pi$

(b) explain why the inverse of f is only defined for $1 - \pi \leqslant x \leqslant \pi - 1$.

The answers are on pages 108–109.

Exponential and logarithm functions

The **exponential function**, $f(x) = e^x$ or $\exp(x)$, can be defined either as a limit or as a power series.

$$e^x = \lim_{n\to\infty}\left(1 + \frac{x}{n}\right)^n \quad \text{or} \quad e^x = 1 + x + \frac{x^2}{2!} + \frac{x^3}{3!} + \ldots$$

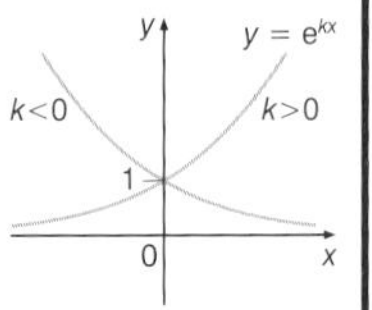

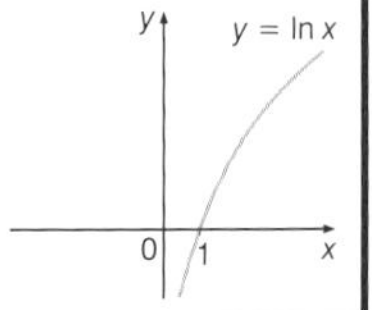

The function e^{kx} models exponential growth if $k > 0$ and exponential decay if $k < 0$.

y
$y = e^{kx}$
$k<0$
$k>0$
1
0
x

The **natural logarithm** $\ln x$ is the inverse function of e^x; if $y = e^x$ then $x = \ln y$. The logarithmic function is undefined for a negative argument, since $e^x \geqslant 0$. Natural logarithms are sometimes called **Napierian logarithms**, after the Scottish mathematician John Napier.

y
$y = \ln x$
0
1
x

From the relationship $y = e^x \Leftrightarrow x = \ln y$ you can see that $\ln y$ can be regarded as a power of e: e is the **base** of the logarithms.

Common logarithms, written as $\log x$, are logarithms to base 10. They obey the same set of rules as natural logarithms; the graph is similar in shape to that of $\ln x$ but with different scales.

Properties of logarithms

These laws, expressed for natural logarithms, hold for logarithms to any base.

- $\ln 1 = 0$
- $\ln x^p = p\ln x$
- $\ln ab = \ln a + \ln b$
- $\ln\frac{a}{b} = \ln a - \ln b$

Q Cars depreciate continuously from when they are bought. The value £V of a car, purchased new for £10 000, after t years is $V = Ae^{-0.25t}$.
(a) Determine the value of A.
(b) What is the value of the car after 5 years?
(c) How long does it take for the value to fall to £6000?

(a) When $t = 0$, $V = Ae^0 = 10\,000 \Rightarrow A = 10\,000$ so $V = 10\,000e^{-0.25t}$.
(b) When $t = 5$, $V = 10\,000e^{-0.25\times 5} = 10\,000e^{-1.25} = 10\,000 \times 0.2865$ $= £2865$.
(c) Suppose it takes T years, then $10\,000e^{-0.25T} = 6000$ $\Rightarrow e^{-0.25T} = 0.6 \Rightarrow -0.25T = \ln 0.6 \Rightarrow T = 2.04$ years.

Q Solve these equations. **(a)** $4\ln x^3 + 5\ln x = 4$ **(b)** $e^{2x} - 3e^x + 2 = 0$

(a) Rewrite the left-hand side as $4 \times 3\ln x + 5\ln x = 4$ (see the properties listed above).
Then $17\ln x = 4 \Rightarrow \ln x = \frac{4}{17} \Rightarrow x = e^{\frac{4}{17}} = 1.265$.

(b) Using the laws of indices the equation can be written as $(e^x)^2 - 3e^x + 2 = 0$ which is a quadratic. Factorising gives $(e^x - 2)(e^x - 1) = 0 \Rightarrow e^x = 1, 2$ (or use the formula). Finally $e^x = 1 \Rightarrow x = \ln 1 = 0$ and $e^x = 2 \Rightarrow x = \ln 2 = 0.693$.

Modelling with logarithms

Taking common logarithms of the data set $x \in [1, 10000]$ gives the set $\log x \in [0, 4]$. Logarithms (common or natural) have a significant scaling effect on data sets. You can use this property to help linearise graphs of data sets and to find their relationship.

If a set of data (x_i, y_i) is believed to obey a relationship $y = ax^b$, plot a graph of $\log y$ vertically against $\log x$ (use common or natural logarithms). This is a log-log plot. If it is a good straight line, the power relationship is obeyed. The values of a and b are given by b = slope of line and $a = 10^{\text{intercept}}$ for common logarithms or $e^{\text{intercept}}$ for natural logarithms.

If the data set is believed to obey a relationship $y = ae^{bx}$, plot a graph of $\ln y$ vertically against x horizontally (a semi-log plot). If it is a good straight line, the exponential relationship is obeyed and b = slope of line, $a = e^{\text{intercept}}$.

Q These displacement–time data are believed to obey a power law $s = at^b$. By plotting a suitable graph, determine the values of a and b.

t (s)	0.5	1.0	1.5	2.0	2.5	3.0
s (m)	0.35	1.4	3.15	5.6	8.75	12.6

Construct a table of values of $\log t$ and $\log s$. Either type of logarithm can be used: in this application natural logarithms have been used.

$\ln t$	−0.69	0	0.41	0.69	0.92	1.10
$\ln s$	−1.05	0.34	1.15	1.72	2.17	2.53

Now plot $\ln s$ vertically against $\ln t$. This gives a good straight line, so the data obey the power law.

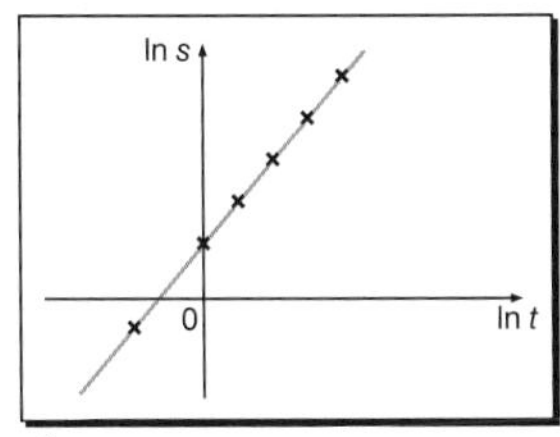

The intercept on the vertical axis $= 0.34 \Rightarrow a = e^{0.34} = 1.4$.

$$b = \text{slope of line} = \frac{2.53 - (-1.05)}{1.10 - (-0.69)} = \frac{3.58}{1.79} = 2.$$

So the data obey the relationship $s = 1.4t^2$.

Note that if the original data set had included $(s, t) = (0, 0)$ then either you could not use the logarithm approach (since $\log 0$ is undefined) or you could discard the (0, 0) data values and perform the analysis on the remaining set of data values. However, always be cautious about discarding awkward data – it just might be the piece that is crucial to the whole analysis.

Exponentials and Logarithms

1 After a patient has been injected with a certain drug the quantity (milligrams) left in the blood t hours later is given by $C = C_0e^{-0.15t}$ where C_0 is a constant.

(a) If the initial dose is 10 milligrams, determine the value of C_0.

(b) Sketch the graph of C against t for $0 \leqslant t \leqslant 6$,

(c) How long does it take for C to reduce to half the initial dose?

2 If $f(x) = e^x$ show that $f(x + y) = f(x).f(y)$ and $(f(x))^n = f(nx)$.

3 Solve these equations.

(a) $\frac{1}{2}(e^x - e^{-x}) = 2$

(b) $\ln(x + 1) - \ln x = 0.1$

4 Use the properties of logarithms to explain how the graph of $y = \ln 3x$ could be obtained from that of $y = \ln x$.

5 The following data are believed to follow an exponential relationship of the form $y = ae^{bx}$.

By plotting a suitable graph, determine the values of a and b.

x	0	5	10	15	20	25
y	2	1.21	0.74	0.45	0.27	0.16

The answers are on page 110.

Interval bisection to solve f(x) = 0

If a function f(x) has a root in the interval $[a, b]$ then the function values f(a) and f(b) are of opposite sign.

You can find the root by first calculating $c = \frac{1}{2}(a + b)$, the midpoint of [a, b] and then f(c). One of the pairs of function values f(a), f(c) or f(b), f(c) will be of opposite signs, which places the root in an interval half the original size. Repeat the halving process as necessary to give the root to the required accuracy.

Q Confirm that the equation $x^3 + x^2 - 3 = 0$ has a root in [1, 1.5]. Use the method of bisection to find the root to within an accuracy of 0.05.

Let $f(x) = x^3 + x^2 - 3$. Then $f(1) = -1$ and $f(1.5) = 2.625$.

There is a sign difference, so the root $\in [1, 1.5]$.

The bisection method can be laid out as follows, changing a and b so that the root always lies in [a, b].

a	f(a)	b	f(b)	c = $\frac{1}{2}$(a + b)	f(c)	|b − a|, interval width
1	−1	1.5	2.625	1.25	0.516	0.5
1	−1	1.25	0.516	1.125	−0.311	0.25
1.125	−0.311	1.25	0.516	1.1875	0.0847	0.125
1.125	−0.311	1.1875	0.0847	1.15625	−0.1173	0.0625

The interval width is now less than 0.1 so the root is known to within 0.05 and is taken as 1.156 25.

Fixed point iteration

Rewrite the equation f(x) = 0 in the form $x = g(x)$. Set up the iterative scheme $x_{n+1} = g(x_n)$ with x_0 as the initial approximation to the root. The sequence of iterates will converge to the root if $|g'(x_0)| < 1$.

You should always use your graphics calculator for iterative methods.

Q The equation $x^3 - 12x - 2 = 0$ has a root in $[-1, 0]$. With $x_0 = 0$ use the method of fixed point iteration to obtain an estimate of the root, accurate to 5 decimal places.

There are many possible rearrangements: two obvious ones are $x = \frac{1}{12}(x^3 - 2)$ and $x = \sqrt[3]{2 + 12x}$.

The first is easier to use, so $g(x) = \frac{1}{12}(x^3 - 2) \Rightarrow g'(x) = \frac{1}{4}x^2$.

Thus $|g'(x_0)| = |g'(0)| = 0 < 1$ so convergence is expected.

The iterative scheme is $x_{n+1} = \frac{1}{12}(x_n^3 - 2)$.

The first few iterates are shown. When two consecutive iterates agree, to 5 decimal places, the process can stop. This gives the root as $-0.167\,06$.

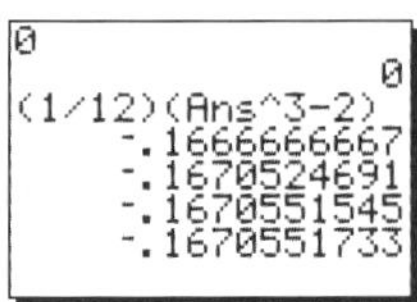

Newton Raphson method

This method uses the iterative scheme $x_{n+1} = x_n - \frac{f(x_n)}{f'(x_n)}$ to solve the equation $f(x) = 0$, with x_0 an initial approximation to the root. Of the three solution techniques examined here the Newton-Raphson method generally converges fastest (unless $f'(x_n) \approx 0$).

Q Use the Newton Raphson method with $x_0 = 0$ to solve the equation $x^3 - 12x - 2 = 0$, to obtain a root which is accurate to 5 decimal places.

$f(x) = x^3 - 12x - 2$ so $f'(x) = 3x^2 - 12$.

The iterative scheme is thus $x_{n+1} = x_n - \frac{x_n^3 - 12x_n - 2}{3x_n^2 - 12}$.

Use a graphics calculator to find the iterates as shown.

Hence the root is $-0.167\,06$ to 5 d.p.

Note that convergence was faster than the fixed point method.

```
0
                0
Ans-(Ans^3-12Ans
-2)/(3Ans²-12)
     -.1666666667
     -.1670551671
     -.1670551734
```

Trapezium rule

This method of integration is based on approximating the area under a curve by a finite number of trapezia (of equal width). The integral is approximated by the sum of areas of trapezia. As the number of trapezia is increased (for a given range of integration) the approximation improves.

$$\int_a^b y\,dx = \frac{h}{2}\{(y_0 + y_n) + 2(y_1 + y_2 + \ldots + y_{n-1})\}$$

$h = \dfrac{b-a}{n}$ where n is the number of trapezia.

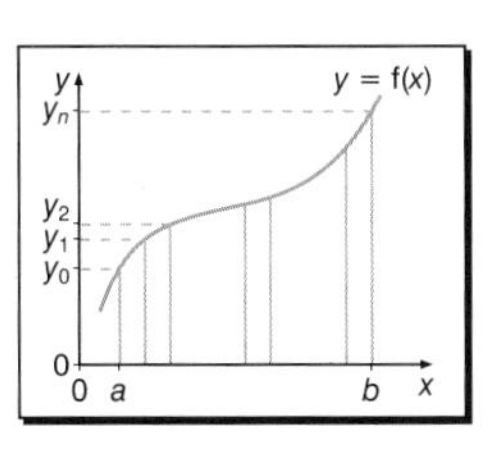

Q The table shows the values of $y = f(x) = \dfrac{4}{1+x^2}$ on the interval [0, 1].
Use these data and the trapezium rule with:
(a) $h = 0.1$ **(b)** $h = 0.2$
to approximate the integral. If it is known that the true value of this integral is π, comment on the results obtained in (a) and (b).

x	0	0.1	0.2	0.3	0.4	0.5	0.6	0.7	0.8	0.9	1
y	4	3.960	3.846	3.670	3.448	3.2	2.941	2.685	2.439	2.210	2

(a) With $h = 0.1$ all the data values are used to give:

$$\int_0^1 y\,dx = \frac{0.1}{2}\{(4 + 2) + 2(3.960 + 3.846 + 3.670 + \ldots + 2.210)\}$$
$$= 3.1399$$

(b) With $h = 0.2$ only data corresponding to $x = 0, 0.2, 0.4, \ldots 1.0$ are used to give:

$$\int_0^1 y\,dx = \frac{0.2}{2}\{(4 + 2) + 2(3.846 + 3.448 + 2.941 + 2.439)\}$$
$$= 3.1348$$

The result corresponding to $h = 0.1$ is closer to the true value, illustrating that the approximation becomes more accurate as the step size is reduced.

Check yourself

Numerical Methods

1 Show that the equation $x^2 - x - 1 = 0$ has a solution in the interval [1.5, 2]. Use the bisection method to determine the solution for which the maximum error is less than 0.05.

2 Show that the equation $x^2 - x - 1 = 0$ can be rewritten as $x = 1 + \frac{1}{x}$.

Use fixed point iteration, with $x_0 = 1.5$, to obtain one root of the quadratic equation to an accuracy of 3 decimal places.

3 Use the Newton Raphson method with $x_0 = 2$ as a first approximation to the root to solve the equation $x^2 - x - 1 = 0$. Continue iterating until the root is obtained to an accuracy of 3 decimal places.

4 Use the trapezium rule to estimate the area under the curve, from $x = 0$ to $x = 10$, given by the following table.

x	0	2	4	6	8	10
y	0	12	48	108	192	300

The answers are on page 111.

Different forces

When you are considering any problem on force, you need to take the following forces into account.

Gravity This is the force that the Earth exerts on all objects. It acts towards the centre of the Earth. If the mass of the object is m kg then the force of gravity is defined to be mg newtons (mg N) where g is the acceleration due to gravity ($g = 9.8\,\mathrm{m\,s^{-2}}$).

Normal reaction If you sit on a chair, gravity exerts a *downward* force on you. The chair exerts an *upward* force on you to balance the force of gravity. This is a normal reaction force. It acts perpendicular to the surface and is usually denoted by R.

Friction Whenever two rough surfaces are in contact, the friction force (F) acts in the direction to oppose motion. $F \leqslant \mu R$ where μ is the **coefficient of friction** and R is the normal reaction force.

Tension and thrust These come into play when objects are attached to rods, strings or springs.

Air resistance This acts in the opposite direction to motion.

Upthrust or buoyancy This occurs when an object is in a fluid. It acts in the opposite sense to gravity.

Force diagram

You should always draw a diagram to show all the forces that are acting. Go through the list of forces above to make sure that you have included everything.

Here are some examples of situations and their force diagrams.

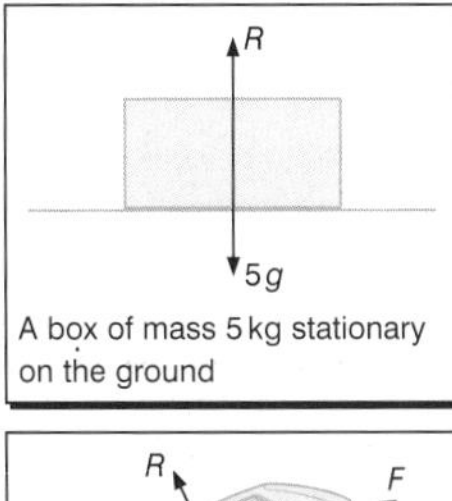

A box of mass 5 kg stationary on the ground

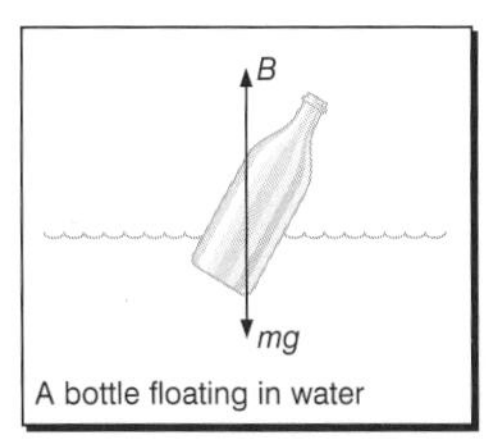

A bottle floating in water

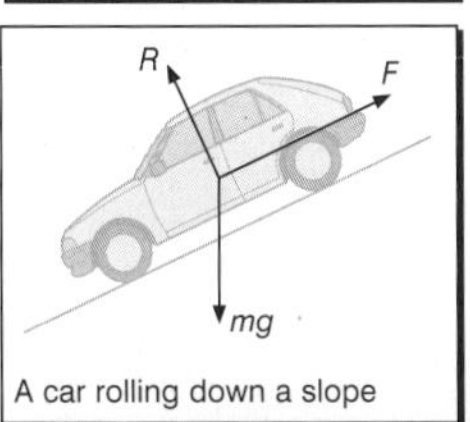

A car rolling down a slope

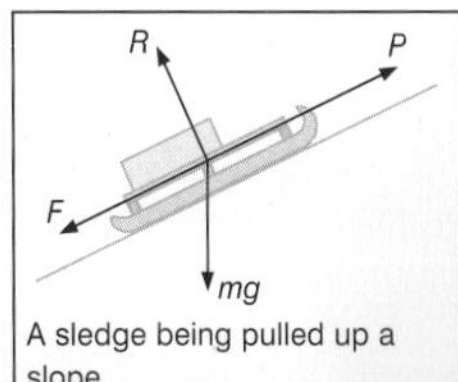

A sledge being pulled up a slope

Balanced and unbalanced forces

In most situations, bodies are subject to more than one force. The overall effect will be determined by the sum (**resultant**) of the forces, taking into account their individual **magnitudes** and **directions**.

- If the resultant is zero, the motion of the object will be unchanged.
- If the resultant is non-zero, the motion of the object will change.

A change in motion can be a change in **velocity** or a change in **direction**.

For example, the forces acting on a free-fall parachutist are the force of **gravity** (downwards) and the **air resistance force** (upwards) to oppose motion.

In the early stages of her freefall the force of gravity is greater than the resistance force. So there is a net downward force which accelerates the parachutist downwards.

As her velocity increases the magnitude of the air resistance increases. If she freefalls for long enough, she will reach a point at which the force of gravity (downwards) is equal to the air resistance (upwards) and the forces balance. At this stage there is no net downward (or upward!) force, so there will be no change to her motion (i.e. no change in her speed). She will continue to fall at this **terminal velocity** until she opens her parachute.

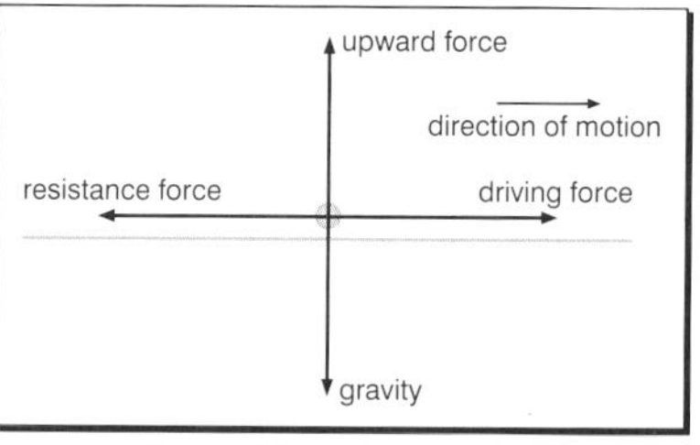

Consider a car cruising along a road at a steady speed. Modelling the car as a particle, the forces on it are in balance, as shown in the diagram.

If the driver decides to speed up, he must **increase** the magnitude of the **driving force**. While the vertical forces still balance each other, the magnitude of the driving force is now greater than that of the **resistance force**. There is a net horizontal force acting to the right, which leads to the change in the motion.

If the car travels round a bend, its **velocity** will change (because the direction of motion is changing even though the speed remains constant). This **change in the motion** of the car is caused by a **change in the forces** acting on the car.

Q Four boxes, each of mass 3 kg, are stacked vertically. Draw a diagram showing the forces acting on each box and find the magnitude of each force.

The figure shows the weights and reaction forces experienced by each box. Since all the boxes are at rest then the net force on each box is zero.

For Box A: $R_1 = 3g$
For Box B: $R_2 = R_1 + 3g = 6g$
For Box C: $R_3 = R_2 + 3g = 9g$
For Box D: $R_4 = R_3 + 3g = 12g$

Q A box of mass m is placed on a slope, inclined at an angle α to the horizontal. It is prevented from moving down the slope by a fixed plate which is perpendicular to the slope and against which the block rests.

Draw a diagram to show all the forces acting on the block and determine the magnitude of each normal reaction.

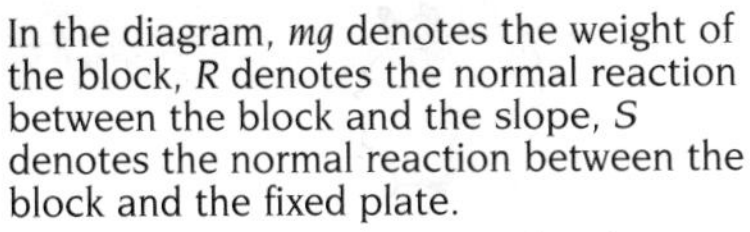

In the diagram, mg denotes the weight of the block, R denotes the normal reaction between the block and the slope, S denotes the normal reaction between the block and the fixed plate.

Since the block is at rest on the slope then the force system is balanced. Resolving the forces on the block in a direction perpendicular to the slope:

$R = mg\cos\alpha$

Resolving the forces on the block in a direction parallel to the slope:

$S = mg\sin\alpha$

Q The diagram shows a system of masses and strings at rest. Draw two separate diagrams to show the forces acting on the two masses. Find the tension in each string.

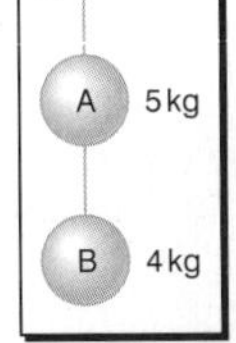

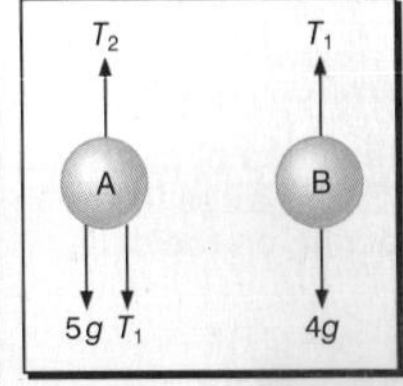

The diagram shows the forces acting on the two masses. Since they are at rest, upward forces equal downward forces and so $T_1 + 5g = T_2$ and $4g = T_1$. Solving gives $T_1 = 39.2$ N and $T_2 = 88.2$ N.

The universal law of gravitation

The universal law of gravitation describes the force of gravitational attraction between any two objects.

The force of attraction between two objects $= \frac{Gm_1m_2}{r^2}$ where m_1 and m_2 are the masses of the objects, G is a constant, known as the **universal gravitational constant** ($6.67 \times 10^{-11}\,\text{kg}^{-1}\,\text{m}^{-3}\,\text{s}^{-2}$) and r is the distance between their centres.

Q What is the magnitude of the force of attraction between a person of mass 60 kg and their partner of mass 70 kg, when they are standing 1 m apart?

$$\text{The force of attraction} = \frac{6.67 \times 10^{-11} \times 60 \times 70}{1^2}$$
$$= 2.8 \times 10^{-7}\,\text{N}$$

This is extremely small and can be ignored.

Expressing forces as vectors

You can use vectors to split a force acting in any direction into two perpendicular components, often taken as a vertical and a horizontal component. The horizontal vector is expressed in terms of **i** and the vertical in terms of **j** where **i** and **j** are perpendicular **unit vectors**.

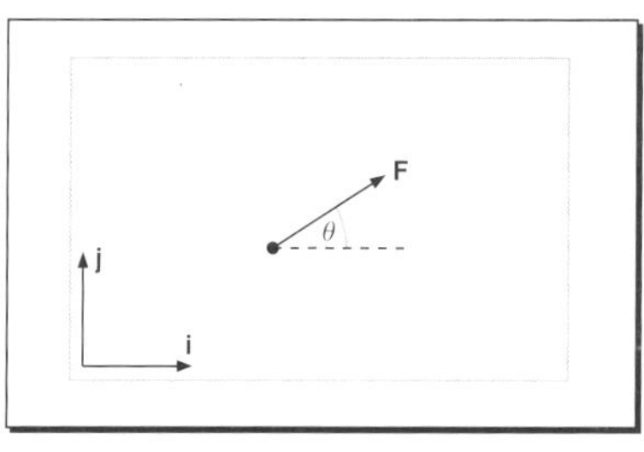

The diagram shows the plan view of a person pushing an object across a smooth floor.

If F denotes the magnitude of the applied force and if unit vectors **i** and **j** are chosen to be parallel to the sides of the room as shown, then the force vector (**F**) can be resolved into components as:

$\mathbf{F} = F\cos\theta\mathbf{i} + F\sin\theta\mathbf{j}$

where $F = |\mathbf{F}|$

Q Express the force illustrated in the diagram in terms of perpendicular unit vectors **i** and **j**.

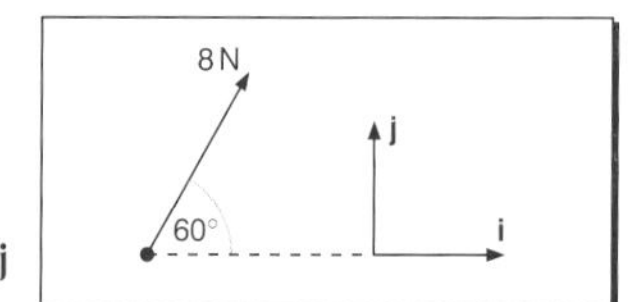

$\mathbf{F} = 8\cos60°\mathbf{i} + 8\sin60°\mathbf{j} = 4\mathbf{i} + 4\sqrt{3}\mathbf{j}$

Q Find the magnitude and direction of the force represented by $-6\mathbf{i} + 8\mathbf{j}$ where $\mathbf{i}$ and $\mathbf{j}$ are perpendicular unit vectors.

$\mathbf{F} = -6\mathbf{i} + 8\mathbf{j}$

So the magnitude $= |\mathbf{F}| = \sqrt{6^2 + 8^2}$
$= \sqrt{100}$
$= 10$

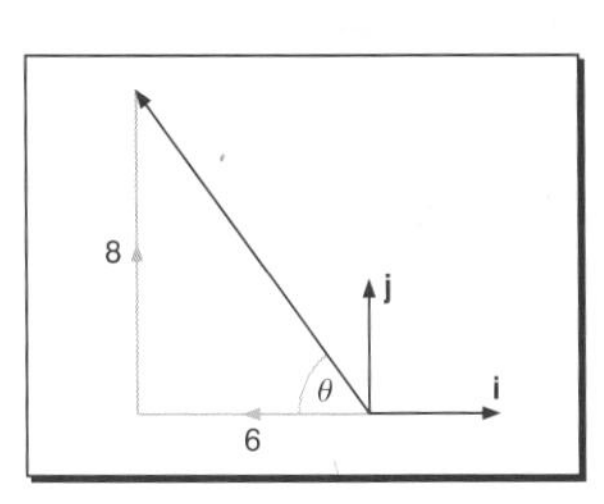

The direction of the force is usually measured from the $\mathbf{i}$-axis, so in this case $\tan\theta = \frac{8}{6}$.

The angle is $\tan^{-1}\frac{8}{6} = 53.13°$

clockwise from the negative $\mathbf{i}$-axis.

Forces and equilibrium

Newton's first law says that if the resultant force on an object is zero then the motion of the body will be unchanged. This means that either the body will remain at rest or it will move with uniform speed in a straight line. You can use forces in their **vector form** to establish whether a body is in equilibrium, by finding the resultant force and checking if it has zero magnitude.

Q A particle which is initially at rest is acted upon by three forces, each acting in the x–y plane.

$\mathbf{F}_1 = 3\sqrt{3}$ N acts in the positive x-direction,
$\mathbf{F}_2 = 3$ N acts in the negative y-direction and
$\mathbf{F}_3 = 6$ N acts in the direction $-\frac{\sqrt{3}}{2}\mathbf{i} + \frac{1}{2}\mathbf{j}$.

Comment upon the motion of the particle.

$\mathbf{F}_1 = 3\sqrt{3}\mathbf{i} + 0\mathbf{j}$
$\mathbf{F}_2 = 0\mathbf{i} - 3\mathbf{j}$
$\mathbf{F}_3 = 6(-\frac{\sqrt{3}}{2}\mathbf{i} + \frac{1}{2}\mathbf{j}) = -3\sqrt{3}\mathbf{i} + 3\mathbf{j}$

By addition, the resultant force $\mathbf{R} = 0\mathbf{i} + 0\mathbf{j} = \mathbf{0}$.

Since the resultant force has zero magnitude, the motion of the particle is unchanged i.e. it remains at rest

Check yourself

Forces

1 A man is trying to stop a packing case sliding down a rough slope inclined at 30° to the horizontal. He is pulling on a rope which is parallel to the slope. If the packing case has a mass of 60 kg and the coefficient of friction between the slope and the case is 0.4, calculate the tension in the rope if the packing case is just on the point of sliding.

2 Draw diagrams showing all the forces acting on a particle of mass m as it is:

(a) pulled up a rough slope inclined at an angle α to the horizontal
(b) allowed to slide down the slope under gravity.

3 The current world record in the high-jump is 2.3 m. If a high-jump competition was held on the moon what would be the equivalent of the Earth record? (Radius of moon $= 1.73 \times 10^6$ m, mass of moon $= 7.38 \times 10^{22}$ kg)

4 **(a)** Find the tension in the string if there is no movement of the system in the diagram.
(b) Find the friction force present between the block and the table.

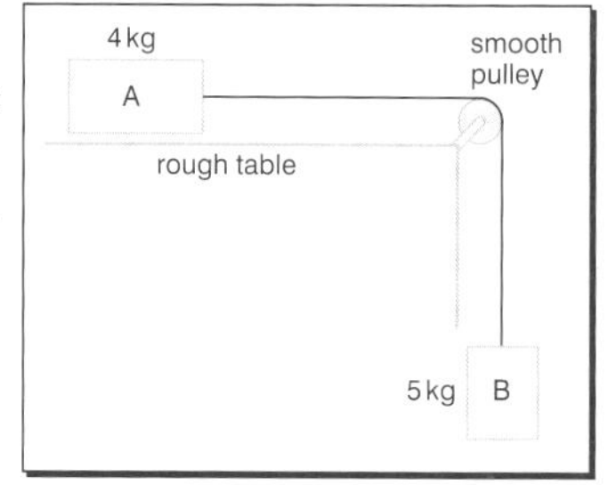

5 A particle of mass 3 kg hangs in equilibrium, supported as shown in the diagram.

Find the tension in each string.

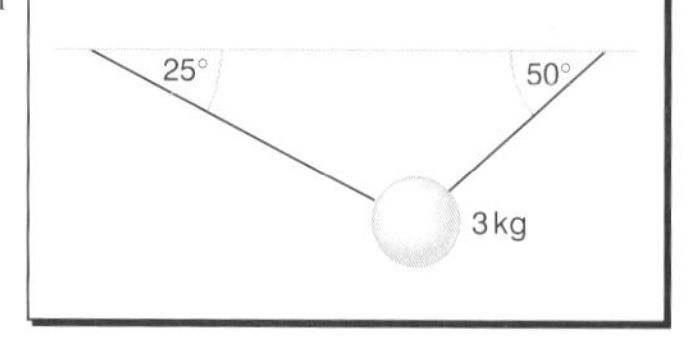

The answers are on page 112.

Forces

6 A box of mass 2 kg is placed on a rough plane inclined at 30° to the horizontal. The box is being pulled up the slope by a force of 19.6 N parallel to the slope. If the box is just about to move, show that the coefficient of friction is $\frac{1}{\sqrt{3}}$.

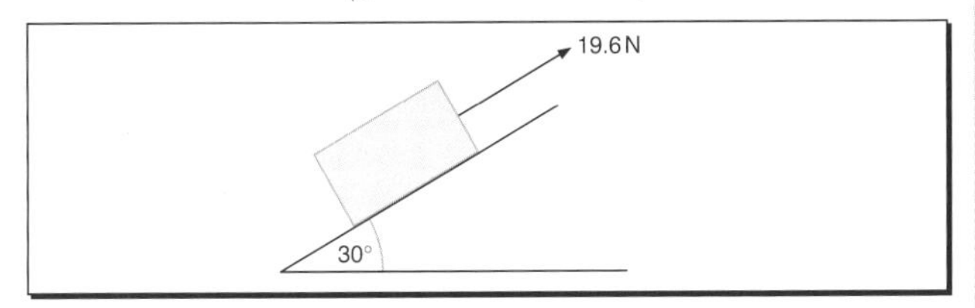

7 The system shown in the diagram is in equilibrium. Find the magnitude and direction of the force **F**.

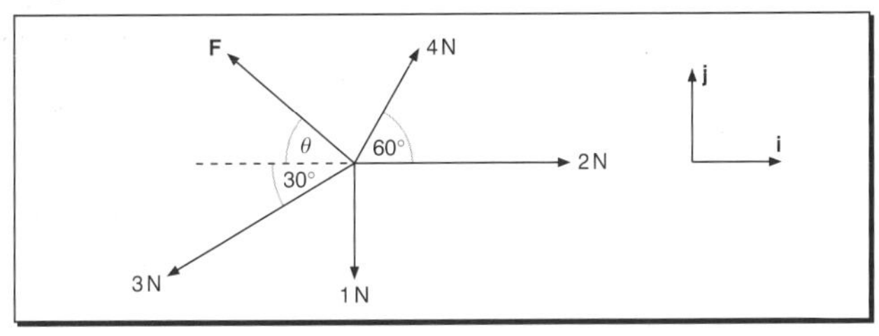

8 A car of mass 1000 kg is being driven along a straight horizontal road.

(a) Draw a diagram showing the forces acting on the car.

(b) What is the relationship between the forward forces and the resistive forces when the car is travelling at its maximum speed?

(c) The resistive forces are related to the speed (v) by $R = v^2 + 20v$. Show that when freewheeling down a slope of 10° the car experiences a forward force parallel to the slope of 1702 N. What is the maximum speed of the car as it freewheels down the slope?

The answers are on page 113.

Uniform motion

Suppose a runner runs along a road at a steady speed of $5\,\mathrm{m\,s^{-1}}$. Since her speed is constant her motion is **uniform**. A graph of her speed (plotted vertically) against time is a horizontal line. The distance travelled between two times, t_1 and t_2, is given by the area shaded in the graph.

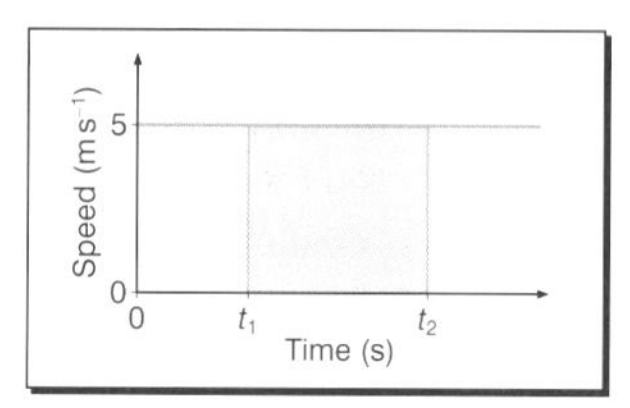

If you calculate the distance travelled between $t = 0$ and, say, $t = 1, 2, 3, \ldots$ seconds and plot these results in a graph of distance travelled against time (plotted horizontally), you obtain a straight line. The equation of the straight line is $s = 5t$. The gradient of the displacement–time graph is equal to the speed of the runner.

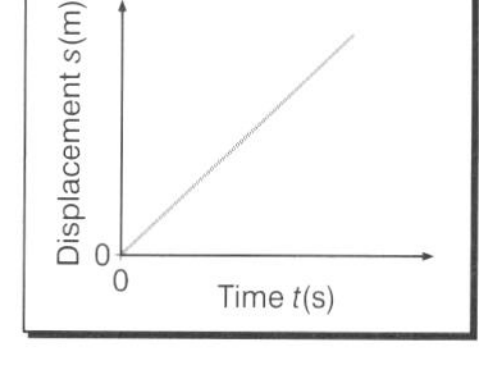

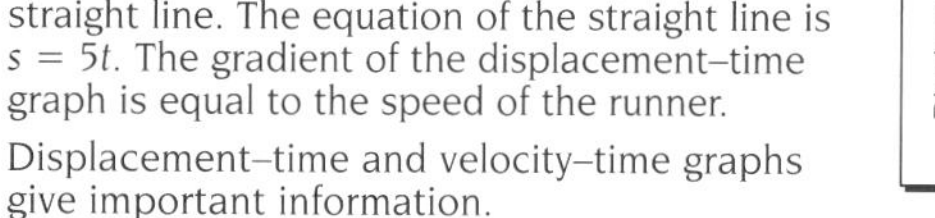

Displacement–time and velocity–time graphs give important information.

- In a **displacement–time** graph, the **gradient** gives the **velocity.**
- In a **velocity–time** graph, the **area under the graph** gives the **distance travelled**, and the **gradient** gives the **acceleration**.

Q The velocity of a 100 m sprinter increases from 0 to $5\,\mathrm{m\,s^{-1}}$ in 1.25 seconds. Assuming that the velocity increases at a constant rate:

(a) sketch a graph to show the velocity against time
(b) interpret the value of the gradient in the context of the sprinter
(c) find the distance travelled by the sprinter between:
  (i) $t = 0$ and $t = 0.5$
  (ii) $t = 0$ and $t = T\ (0 < T \leqslant 1.25)$.

(a) You can join the points (0, 0) and (1.25, 5) by a straight line because the increase in velocity is assumed to be uniform. The gradient of the line is $\frac{5}{1.25} = 4$,

so the equation of the line is $v = 4t$.

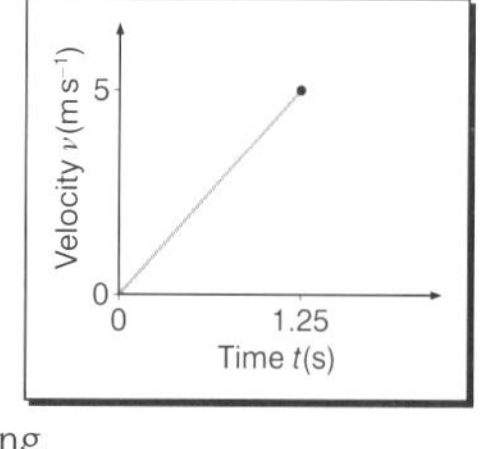

(b) The gradient measures the rate of change of velocity with respect to time which is the acceleration. The acceleration of the sprinter is $4\,\mathrm{m\,s^{-2}}$ during the first 1.25 seconds of her running.

(c) (i) The distance travelled between $t = 0$ and $t = 0.5$ s is given by the area indicated on the velocity–time graph.
Distance travelled between $t = 0$ and $t = 0.5$ s
$= \frac{1}{2} \times 0.5 \times (4 \times 0.5) = 0.5$ m

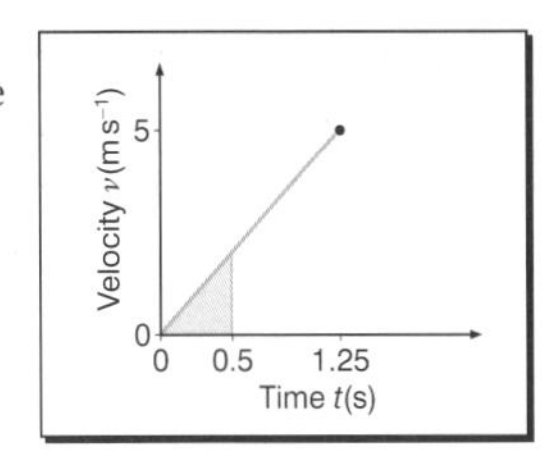

(ii) Similarly the distance travelled between $t = 0$ and $t = T$ is given by the area shown.

Distance travelled between $t = 0$ and $t = T$
$= \frac{1}{2} \times T \times (4 \times T) = 2T^2$ m

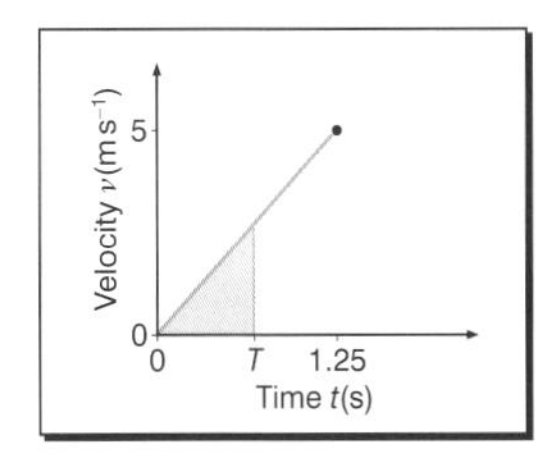

Constant acceleration

The motion of a body moving with constant acceleration is described by the three equations:

- $v = u + at$
- $v^2 = u^2 + 2a(s - s_0)$
- $s = s_0 + ut + \frac{1}{2}at^2$

where a (in m s^{-2}) denotes the value of the constant acceleration, t (in seconds) denotes time, s (in metres) is the displacement of the body at time t, s_0 (in metres) is the initial displacement of the body, u (in m s^{-1}) is the initial velocity of the body and v (in m s^{-1}) is the velocity of the body at time t.

Q A ball is thrown vertically upwards, from a point 1 m above the ground, with a velocity of 5 m s^{-1}. Find:
(a) the maximum height reached above ground level
(b) the time taken to reach this point.

(a) At maximum height, $s_0 = 1$, $s = ?$.
The equation to use is $v^2 = u^2 + 2a(s - s_0)$.
$0 = 5^2 + 2(-9.8)(s - 1)$
Solving for s gives $s = 2.28$ m.
(b) The equation to use is $v = u + at$.
$0 = 5 - 9.8t$
so $t = 0.51$ s.

Check yourself

Kinematics in one Dimension

1 Describe the motion of the particle shown in the velocity–time graph. Find the initial velocity, the final velocity, the maximum velocity and the distance travelled by the particle.

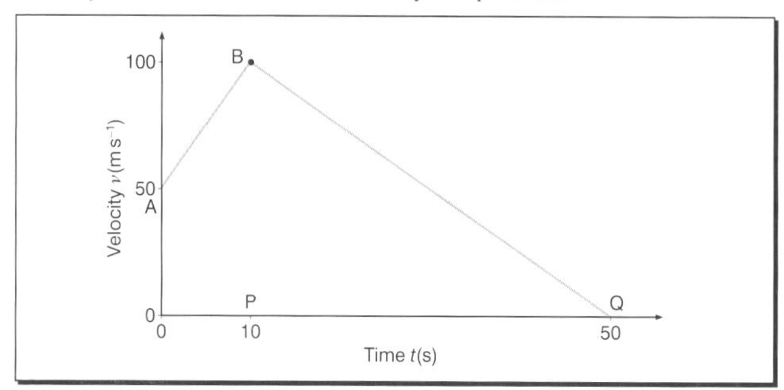

2 A cyclist increases his speed from 0 to 6 m s^{-1} over 12 seconds.
 (a) Sketch the velocity–time graph and hence find the acceleration of the cyclist.
 (b) Find the distance travelled by the cyclist in:
 (i) 4 seconds **(ii)** 8 seconds **(iii)** t seconds ($t \leqslant 12$).
 (c) Assuming that at the end of the 12 seconds he continues travelling, at constant speed, find the distance travelled after 1 minute.

3 A particle moves down a smooth slope inclined at 30° to the horizontal. If it starts from rest, determine its velocity t seconds later and comment upon the realism of the result.

4 A car accelerates at 0.5 m s^{-2} from rest until it reaches a velocity of 20 m s^{-1}. It then applies its brakes and stops after 120 m. What is the total distance travelled by the car and how long was it in motion?

5 A skier pushes off from the top of the ski run with an initial velocity of 1.5 m s^{-1}. He gains speed at a constant rate throughout the run and after 10 seconds he is moving at 5 m s^{-1}. If the length of the ski slope is 400 m, what is his speed at the finish?

The answers are on page 114.

Position vectors

You can describe the motion of an object across a plane in terms of its position **relative to a fixed origin**. You can use Cartesian coordinates to identify the position uniquely, but it is more usual to specify the **distance** and the **direction** from the origin. The position vector is thus expressed in terms of two unit vectors, **i** and **j**, which are perpendicular to each other.

The position of the yacht in the diagram is six kilometres from O on a bearing of 050°.

Its position vector, **r**, is:

$\mathbf{r} = 6\cos 40°\mathbf{i} + 6\cos 50°\mathbf{j}$
$\quad = 4.60\mathbf{i} + 3.86\mathbf{j}$

If the yacht is moving, express each component of the position vector in terms of time-dependent variables x and y:

$\mathbf{r} = x(t)\mathbf{i} + y(t)\mathbf{j}$

To find the path evaluate the position vector at different times and plot the coordinates.

Q The position vector of a snooker ball as it moves across the table is:

$\mathbf{r} = (0.2 + 0.9t)\mathbf{i} + (0.3 + 0.35t)\mathbf{j}$ metres where the origin O is one corner of the table and the **i** and **j** vectors are directed along two edges. Plot the path of the ball between the times $t = 0$ and $t = 2$ s, using a time step of 0.25 s.

At $t = 0$: $\mathbf{r}_0 = 0.2\mathbf{i} + 0.3\mathbf{j}$
At $t = 0.25$s: $\mathbf{r}_1 = 0.425\mathbf{i} + 0.3875\mathbf{j}$ …
At $t = 2$ s: $\mathbf{r}_8 = 2\mathbf{i} + \mathbf{j}$

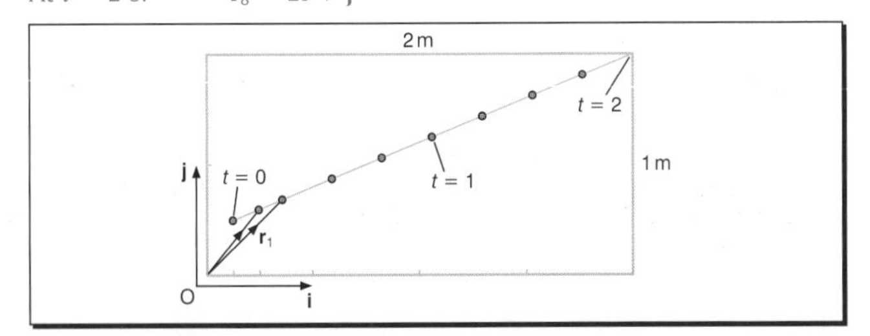

Speed and velocity

Velocity can also be expressed in vector form:

$\mathbf{v} = u\mathbf{i} + v\mathbf{j}$

The speed of an object is the magnitude of the velocity and is a scalar quantity given by:

$|\mathbf{v}| = \sqrt{u^2 + v^2}$

Q A train travelling up a steady incline has a velocity of $35\mathbf{i} + 3\mathbf{j}$ where $\mathbf{i}$ and $\mathbf{j}$ are horizontal and vertical unit vectors respectively. Find the speed of the train and the slope of the incline.

Speed $= \sqrt{35^2 + 3^2} = 35.13\,\text{m s}^{-1}$

The slope of the incline can be found from:

$\alpha = \tan^{-1}\frac{3}{35} = 4.9°$.

Acceleration

Like position and velocity, acceleration can also be expressed as a vector quantity.

If the acceleration is constant, the equations of constantly accelerated motion can be extended to vector form as:

$\mathbf{v} = \mathbf{u} + \mathbf{a}t$ and $\mathbf{r} = \mathbf{r}_0 + \mathbf{u}t + \frac{1}{2}\mathbf{a}t^2$

where $\mathbf{r}_0$ is the position vector for the initial position.

Q A pistol is fired horizontally and the bullet leaves it at $70\,\text{m s}^{-1}$, travelling horizontally. Subsequently the bullet experiences a downward acceleration of $10\,\text{m s}^{-2}$. If the bullet was fired at a height of 1.5 m above ground level, determine when the bullet hits the ground and the distance it has travelled horizontally.

In terms of the vectors $\mathbf{i}$ and $\mathbf{j}$, $\mathbf{u} = 70\mathbf{i}$, $\mathbf{a} = -10\mathbf{j}$ and $\mathbf{r}_0 = 1.5\mathbf{j}$.

Using $\mathbf{r} = \mathbf{r}_0 + \mathbf{u}t + \frac{1}{2}\mathbf{a}t^2$ gives:

$\mathbf{r} = 70t\mathbf{i} + (1.5 - 5t^2)\mathbf{j}$

The bullet hits the ground when the $\mathbf{j}$-component of $\mathbf{r}$ is zero, which is when:

$1.5 - 5t^2 = 0$ so $t = 0.55$ seconds.

The horizontal distance travelled by the bullet is given by the $\mathbf{i}$-component, $70t$ so when $t = 0.55$, the horizontal distance travelled is 38.5 m.

Check yourself

Motion and Vectors

1 A child on a slide experiences an acceleration of $1.5\mathbf{i} - 2\mathbf{j}$. Assume that the child starts from rest. Find an expression for the position of the child at time t.

The child takes 3 seconds to reach the bottom of the slide. Find the length of the slide and the speed of the child at the bottom.

2 A squash ball hits the front wall, moving at $6\,\mathrm{m\,s^{-1}}$, and rebounds at $5\,\mathrm{m\,s^{-1}}$. The diagram shows the change in direction.

Express both velocities in terms of the unit vectors **i** and **j**. Find the acceleration of the ball while it is in contact with the wall, assuming that this contact is maintained for 0.1 seconds.

5 m s⁻¹
60°
45°
6 m s⁻¹
j
i

3 A speedboat leaves the jetty where it was at rest and experiences an acceleration of $1.5\,\mathrm{m\,s^{-2}}$ on a bearing of 50° for 10 seconds. It then travels with constant velocity on that bearing for a further 30 seconds.

Find how far the boat has travelled:

(a) when it stops accelerating

(b) at the end of 40 seconds.

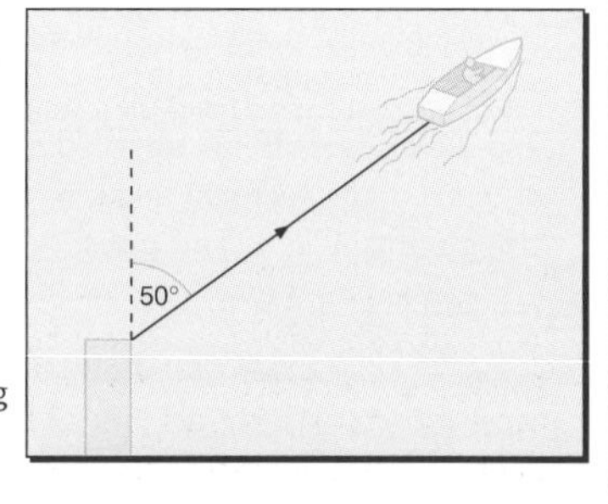

The answers are on page 115.

Newton's first and second laws of motion

If the forces acting on an object are balanced, the **resultant force** on the body is zero. Since forces cause the motion of an object to change, if there is no force acting on the object there will be no change in its motion. Newton's first law of motion is a formal statement of this.

All bodies continue to remain at rest, or to move uniformly in a straight line, unless acted upon by an external force.

If the forces acting on an object are unbalanced, their effect causes a change to the motion of the object. This change is measured in terms of the acceleration **a** given to the object by the force **F**. Newton's second law of motion is a formal statement of this.

A force acting on a body produces an acceleration **a** which is proportional to the applied force **F**. In the case that the mass of the body is m then $\mathbf{F} = m\mathbf{a}$.

Q A car of mass 1000 kg travels down a slope at 10° to the horizontal at a constant speed. A resistance force of 150 N acts parallel to the slope. Find the force exerted by the brakes of the car.

If B represents the braking force then, since the car's motion is uniform, the total force acting up the slope must balance the force acting down the slope.

$B + 150 = 1000g\sin 10°$ so $B = 1552$ N.

Q A ball of mass 300 g, travelling with a speed of $4\,\text{m s}^{-1}$ is struck by a bat. It is deflected by 60° from its original path and its speed is increased to $6\,\text{m s}^{-1}$.

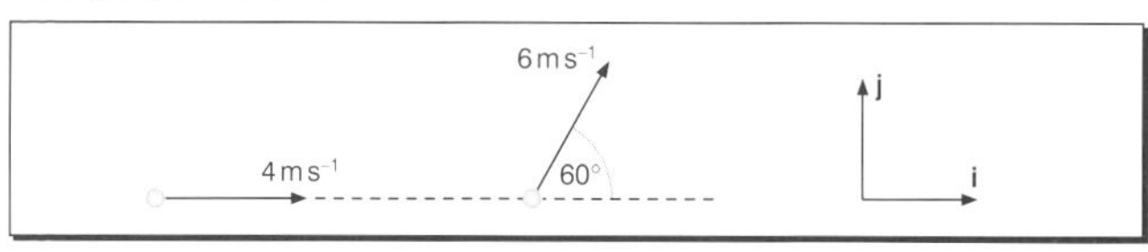

The ball is in contact with the bat for 0.2 seconds. Find the average acceleration of the ball and the average force on the ball.

Change in velocity $= (6\cos 60°\mathbf{i} + 6\sin 60°\mathbf{j}) - 4\mathbf{i} = -\mathbf{i} + 5.196\mathbf{j}$

Average acceleration $= \dfrac{\text{change in velocity}}{\text{time taken}} = -5\mathbf{i} + 25.98\mathbf{j}\,\text{m s}^{-2}$

Using Newton's second law of motion:
average force = mass × average acceleration $= 0.3 \times (-5\mathbf{i} + 25.98\mathbf{j})$
$= -1.5\mathbf{i} + 7.794\mathbf{j}$ N

Connected particles

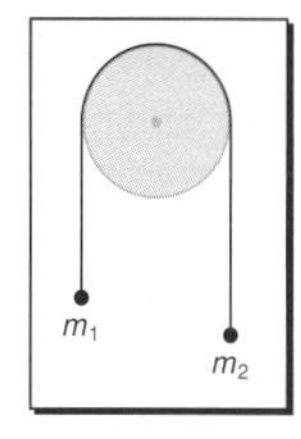

The diagram shows two particles connected by a light string which passes over a smooth pulley.

Provided that the string remains taut, the **tension** in the string on either side of the pulley is the same. Also, the velocity and the acceleration have the same magnitude for both particles.

Q In the system shown in the diagram, the coefficient of friction between the slope and the 2 kg mass is 0.7 and the pulley is smooth. Determine the tension in the string and the acceleration of each mass.

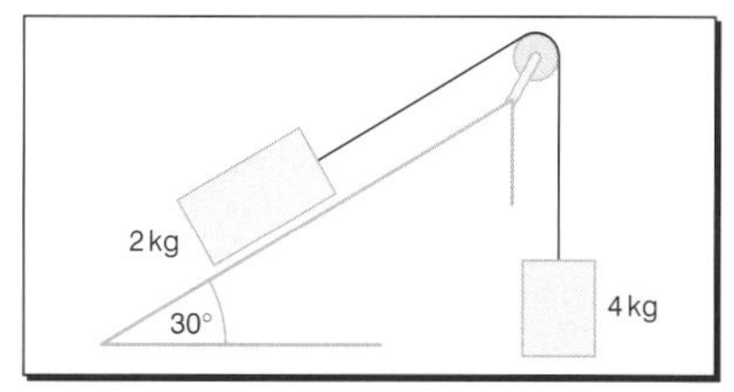

Model the two masses as particles. The forces acting on the two particles are then as shown (where it has been assumed that the 2 kg mass accelerates up the slope with acceleration a).

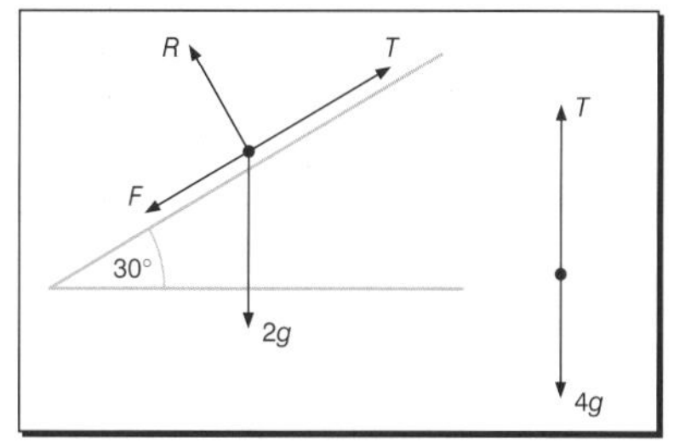

Since the block is sliding on the slope, the law of friction gives:

$F = \mu R = \mu \times 2g\cos30^\circ = 0.7 \times 2g\cos30^\circ$

Apply Newton's second law of motion. For the 2 kg mass as it moves up the slope:

$T - F - 2g\sin30^\circ = 2a$

For the 4 kg mass as it descends vertically:

$4g - T = 4a$

So:

$$\left.\begin{array}{l} T - 21.682 = 2a \\ 39.2 - T = 4a \end{array}\right\} \Rightarrow \begin{array}{l} T = 27.52 \text{ N and} \\ a = 2.92\,\text{m}\,\text{s}^{-2} \end{array}$$

Newton's Laws

1 A skier of mass 70 kg is skiing down a slope at 20° to the horizontal. If the friction force acting on the skier is 100 N, determine her acceleration down the slope.

2 A car of mass 500 kg travelling on a horizontal road is approaching a slope of 1 in 5. When it is 200 m from the foot of the slope its speed is $10\,m\,s^{-1}$. If the maximum drive force the car can produce is 400 N, how far up the slope will it travel, assuming a constant resistive force of 40 N?

3 A particle of mass 2 kg is held on a rough slope which is inclined at 30° to the horizontal. The coefficient of friction between the particle and the slope is 0.4. The particle is attached to a light string which passes over a smooth pulley at the top of the slope. A second particle, of mass 4 kg, is attached to the free end of the string and is hanging vertically under gravity. The system is then released.

Draw a diagram showing all the forces acting on the particles and determine the tension in the string and the acceleration of the system.

4 A man of mass 60 kg is standing in a lift which at a particular moment has an acceleration of $1\,m\,s^{-2}$ upwards. The man is holding a package of mass 5 kg by a single string.

Draw a diagram showing the forces acting on the package, and the direction of acceleration. Show that the tension in the string is 54 N and calculate the reaction of the lift floor on the man.

The answers are on page 116.

Momentum

Momentum is defined as the product of the mass m kg of an object and its velocity, $\mathbf{v}$ m s^{-1}. It is denoted by the vector $\mathbf{p}$.

$\mathbf{p} = m\mathbf{v}$

The study of **collisions and impacts** between objects is based on the change in momentum caused by the impact. This change in momentum is a vector quantity called the **impulse**, $\mathbf{I}$ N s.

Impulse = change in momentum = $m\mathbf{v} - m\mathbf{u}$

where $\mathbf{u}$ is the velocity of the object before the impact and $\mathbf{v}$ is its velocity after the impact.

The impulse can be expressed in terms of the contact force, $\mathbf{F}$ N, acting on the body and the contact time, t seconds. If $\mathbf{F}$ is assumed to be constant then $\mathbf{I} = \mathbf{F}t$.

Conservation of momentum

If no external forces act during the collision, the momentum of the system is conserved. This result is known as the **principle of conservation of momentum**, expressed as:

$m_A\mathbf{u}_A + m_B\mathbf{u}_B = m_A\mathbf{v}_A + m_B\mathbf{v}_B$

where m_A and m_B are the masses of the two bodies A and B, $\mathbf{u}_A$ and $\mathbf{u}_B$ are their respective velocities *before* the collision and $\mathbf{v}_A$ and $\mathbf{v}_B$ are their respective velocities *after* the collision.

Q A model railway truck of mass 0.25 kg is travelling at 2.5 m s^{-1} towards a stationary truck of mass 0.2 kg. Calculate the velocity with which the two trucks move off together, assuming that on collision they join together.

Using conservation of momentum:

$m_A\mathbf{u}_A + m_B\mathbf{u}_B = m_A\mathbf{v}_A + m_B\mathbf{v}_B$

$0.25 \times 2.5\mathbf{i} + 0.2 \times 0\mathbf{i} = 0.45 \times \mathbf{v}$

So $v = 1.39$ m s^{-1}.

Moments

Depending on **where** a force is applied to a body, the resulting motion may be a **rotation**. The rotational effect of a force is measured by its **moment**.

The magnitude M of the moment about O of a force of magnitude F applied to an object at some point P (where OP = d) is defined as

$M = F \times d\sin\theta$

where $d\sin\theta$ is the perpendicular distance from O to the line of action of the force.

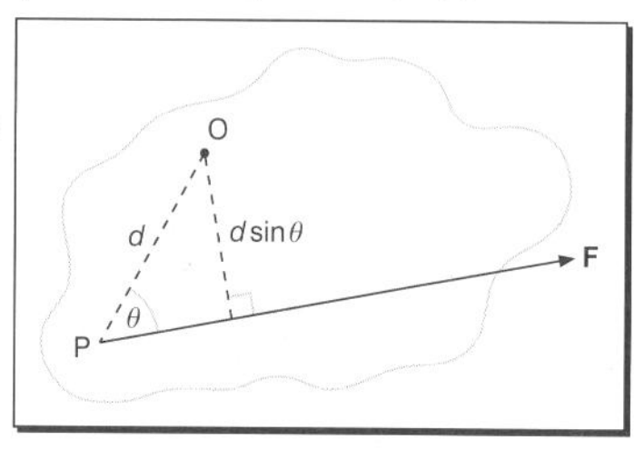

The moment of a force is a vector as it has a direction relative to the axis of rotation. In the diagram, F causes an anticlockwise rotation about O. By convention, anticlockwise is positive.

Conditions for a body to be in equilibrium

For a body to remain in equilibrium, the resultant force on the body and the total moment acting on the body must both be zero.

Q Jenna and Sam sit on a seesaw. Sam has a mass of 30 kg while Jenna's mass is 25 kg.

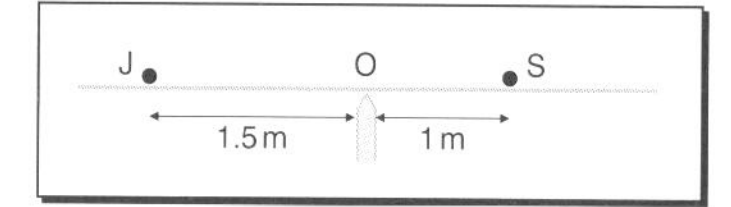

(a) Draw a diagram to show the forces acting on the seesaw and by taking moments about O describe what happens when the seesaw is allowed to move.

(b) Where should Sam sit in order for the seesaw to balance?

(a) Taking moments about O:

Force	Moment (Nm)
J	$25g \times 1.5 = 37.5g = 367.5$
R	0
S	$-30g \times 1 = -30g = -294$
Total	73.5

Sam will rise as the moment is positive i.e. will rotate anticlockwise.

(b) For the seesaw to balance, the total moment must be zero. Let Sam be d m from the pivot. Then $25g \times 1.5 - 30g \times d = 0$ giving $d = 1.25$ m. So Sam must sit 1.25 m from the pivot.

Check yourself

Momentum, Collisions, Moments

1 A ball of mass 200 g travelling horizontally at $6\,\text{m s}^{-1}$ strikes a vertical wall at right angles and rebounds with a speed of $4\,\text{m s}^{-1}$. Find the magnitude of the impulse given to the ball.
Assuming that contact between the ball and the wall lasts for 0.1 seconds, find the force (assumed constant) experienced by the ball.

2 A bullet of mass 30 g is fired horizontally into a block of wood of mass 8 kg. The bullet remains embedded in the block and together they move along a straight line with a velocity of $5\,\text{m s}^{-1}$. What was the velocity of the bullet?

3 A uniform plank AB, of mass 40 kg and length 6 m, rests horizontally on two supports C and D, AC = 2 m and CD = 3 m.

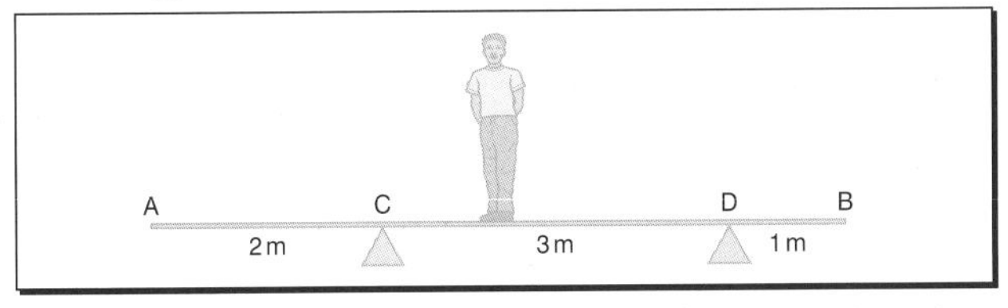

(a) A man weighing $80g$ N stands at the centre of the plank. Find the reactions at C and D.
(b) How far can the man walk towards the end B without overturning it?

4 A uniform ladder of length $2l$ m and mass m kg rests with its top against a smooth vertical wall. The foot of the ladder rests on rough horizontal ground. The coefficient of friction between the ground and the ladder is μ.
If the ladder is on the point of slipping show that $\tan\theta = \dfrac{1}{2\mu}$.

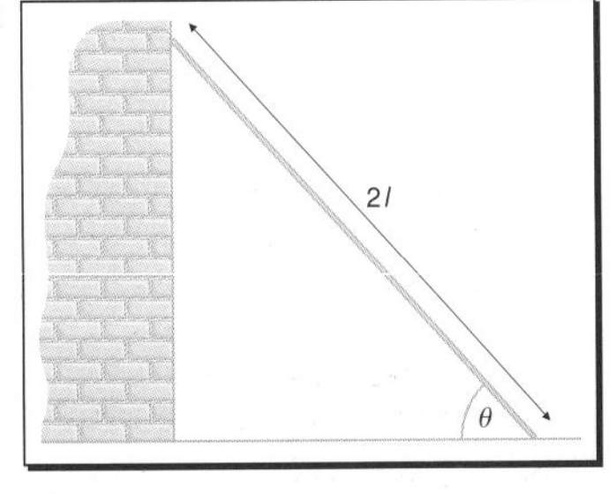

The answers are on page 117.

If you ignore air resistance and spin, the only force acting on a projectile is its **weight** mg.

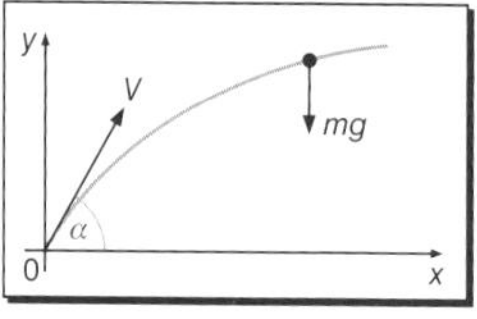

Since this acts vertically downwards, the projectile experiences no acceleration in the horizontal x-direction and an acceleration of $-g$ in the upwards vertical y-direction. Consequently the motion in the x-direction is uniform.

The horizontal and vertical components of the release velocity $V \text{ m s}^{-1}$ are respectively $V\cos\alpha$ and $V\sin\alpha$. If t is the time since the projectile was released then $x = (V\cos\alpha)t$ and $y = (V\sin\alpha)t - 0.5gt^2$.

The time of flight of the projectile is given by:

$$t = \frac{2V\sin\alpha}{g}$$

The range of the projectile is $\dfrac{V^2\sin 2\alpha}{g}$.

The range will be a maximum when $\alpha = 45°$.

e.g. A tennis player serves a ball at a speed of 20 m s^{-1} from an initial height of 2 m. The baseline is 12 m from the net which is 0.9 m high. Find the equation of the path of the ball in terms of x and α, the angle of release. Find the two possible values of α for which the ball just clears the net and explain which value you would recommend.

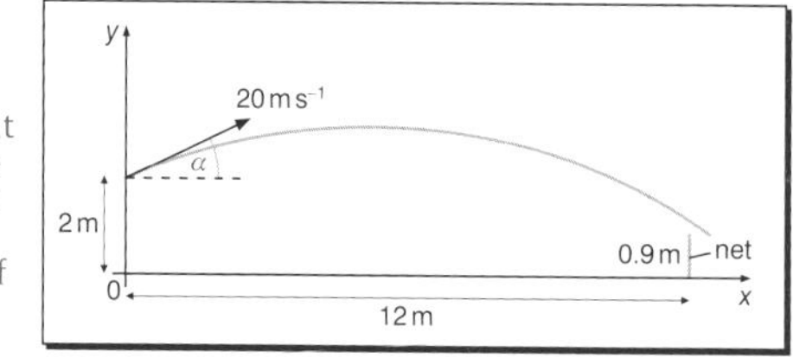

The position of the ball is given by:

$x = (20\cos\alpha)t,\ y = 2 + (20\sin\alpha)t - 4.9t^2$

Eliminating t from these equations gives:

$y = 2 + x\tan\alpha - 0.01225x^2\sec^2\alpha$

For the ball just to clear the net, $y = 0.9$ when $x = 12 \Rightarrow 0.9 = 2 + 12\tan\alpha - 1.764\sec^2\alpha$.

Substituting $\sec^2\alpha = 1 + \tan^2\alpha$ gives a quadratic in $\tan\alpha$:
$1.764\tan^2\alpha - 12\tan\alpha + 0.664 = 0$

Solving the quadratic gives $\tan\alpha = 6.7469$ and $\tan\alpha = 0.0558 \Rightarrow \alpha = 81.6°$ and $\alpha = 3.2°$.

The smaller value of α is recommended because the ball will be in the air for less time, thus giving the opponent less opportunity to decide how to return it.

Check yourself

Projectiles

1. A projectile is launched with a speed of $30\,\text{m s}^{-1}$ in a direction making an angle of 30° with the horizontal. Calculate its position, relative to its starting position, and the magnitude of its velocity after 2 seconds. Comment on the sign of the vertical component of the velocity when $t = 2$.

2. A bullet is projected at a speed of $350\,\text{m s}^{-1}$ at an angle of projection of $\tan^{-1}\frac{3}{4}$.

   Find its speed and direction when it reaches a height of 1250 m relative to its starting position.

3. Show that the maximum height reached by a projectile fired from ground level with a velocity of $v\,\text{m s}^{-1}$ at an angle of α to the horizontal is given by:

   $$\frac{v^2\sin^2\alpha}{2g}$$

4. A particle is fired from the top of a cliff of height 49 m with a speed of $14\,\text{m s}^{-1}$ at an angle of 45° to the horizontal. Find the distance from the foot of the cliff to the point where the particle lands in the sea.

5. A cricket ball is thrown from ground level in the outfield and lands at the feet of the wicket-keeper 60 m away. Show that if u and v are the horizontal and vertical components of the velocity, then $uv = 294$.

The answers are on page 118.

Data

The data collected for statistical analyses can be:

- **qualitative** – **descriptive** information (e.g. colour, gender, race, …)
- **quantitative** – data collected by **counting** (e.g. votes, number of children, …) or **measuring** (e.g. height, weight, wages, …)

Quantitative data can be either **discrete** (integer values only e.g. number of votes) or **continuous** (any values within a given range are possible, e.g. heights).

Different types of data require different types of analysis. Data are often recorded in **frequency tables**. The **frequency** means the number of occurrences of an item or value in a data set (e.g. the number of children who are 1.25 m tall). The frequency values provide information about how the data are distributed.

Displaying data

Data can be displayed either in tabular form (a frequency table) or pictorially (e.g. pie chart, frequency line graph, bar chart, stem and leaf displays).

Q The numbers of different coloured sweets in a packet are shown in the following table. Display the data as a pie chart and a bar chart.

Colour	yellow	green	blue	red	orange
Number of sweets	4	5	3	2	6

Pie chart: The whole set of data is represented by a circle. Each colour is represented by a sector of the circle. The angle of each sector is proportional to the frequency (i.e. the angle of the yellow sector = $\frac{4}{20} \times 360° = 72°$).

Bar chart: The height of each bar represents the frequency of the colour it represents. The bars have gaps between them.

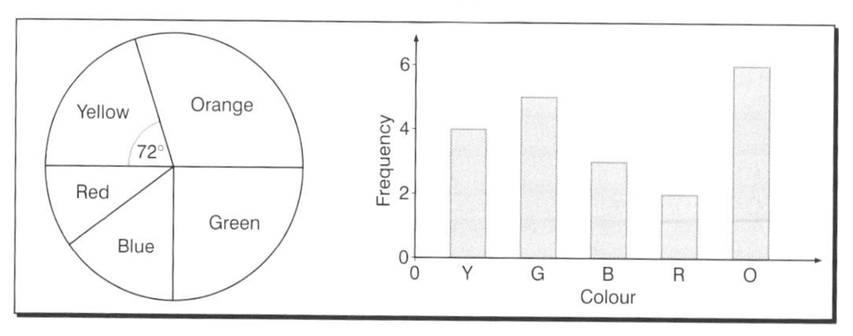

Stem-and-leaf displays

These are used for quantitative data, using the actual data values. The **stem** is the most significant part of the data (e.g. if the data value is 324, the stem is the 3) and the **leaf** is the next most significant part (in this case the 2). Further digits are lost. You must state the unit of the display (here unit = 10, indicating that the 32 actually represents $32 \times 10 = 320$). This two-branch display (so called because two digits are used) includes values in the range 300–399.

Q Display these data in a two-branch stem-and-leaf display.

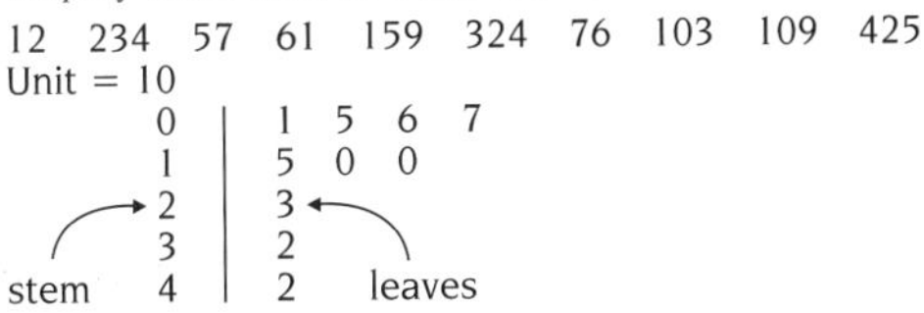

Histograms

These are used to display continuous data graphically. The area of each rectangle is proportional to the frequency of the class it represents. The height of each rectangle is given by the **frequency density** (class frequency ÷ class width). Construct the histogram by plotting frequency density (vertically) against the data classes, x.

Q Display these data as a histogram.

Class	$0 \leqslant x < 1.5$	$1.5 \leqslant x < 2.5$	$2.5 \leqslant x < 4.5$	$4.5 \leqslant x < 10$
Frequency	12	20	32	11

The class widths are different, so find the width of each class, then the frequency density.

Class	**Frequency**	**Class width**	**Frequency density**
$0 \leqslant x < 1.5$	12	1.5	8
$1.5 \leqslant x < 2.5$	20	1	20
$2.5 \leqslant x < 4.5$	32	2	16
$4.5 \leqslant x < 10$	11	5.5	2

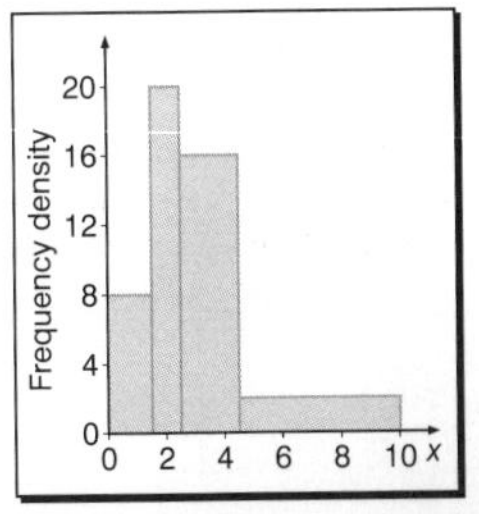

Draw the histogram by plotting frequency density against the variable x.

Summary statistics

You can summarise sets of data in terms of their **location** (measured by the mean, median, mode, weighted mean) and their **spread** (measured by range, interquartile range, variance and standard deviation). For a complete picture of a set of data, you need one measure of location and one of spread.

Measures of location

- **Mean:** the average value of a data set $\{x_i;\ i = 1, 2, \ldots,\}$

  $$\bar{x} = \frac{\Sigma f_i x_i}{n}$$

  where f_i is the frequency of x_i
- **Median:** the middle data value when the n data items are ranked in ascending order i.e. the $\frac{n+1}{2}$th value
- **Mode:** the item with the greatest frequency
- **Weighted mean:** the mean value when different data values (x_i) have different weights (importance) w_i assigned to them

  $$\bar{x}_w = \frac{\Sigma w_i x_i}{\Sigma w_i}$$

Measures of spread

- **Range:** the difference between the largest and smallest data values
- **Interquartile range (IQR):** the range of the middle half of the data
  The items of data that are located midway between the median and the extremes are the **quartiles**, specifically the lower quartile (LQ) and upper quartile (UQ). The difference between the UQ and the LQ is the IQR or **quartile spread** (QS). If there are n items of data then LQ $= \frac{1}{4}(n+1)$th item and UQ $= \frac{3}{4}(n+1)$th item.
- **Standard deviation:**
  This takes all the data into account and is linked to the mean value. The standard deviation of a population is denoted by σ and of a sample is denoted by s. It is calculated using

  $$\sigma = \sqrt{\frac{\Sigma x^2}{n} - \bar{x}^2}$$
- **Variance:** the square of the standard deviation

Use the statistical functions on your calculator to calculate the mean and standard deviation or variance.

Summary statistics for grouped frequency

Data are often presented in frequency tables. So that you can calculate measures of location and spread for such data, you assume that all the data items in each class are equal to the class midpoint. Alternatively, you can describe the distribution of data in terms of the upper class boundaries and a cumulative frequency, for example, 'there are 12 data items < 1.5, 32 data items < 2.5' and so on. Find the median by plotting cumulative frequency against data values.

Q Find the mean, standard deviation, median and IQR of these data.

Class	$0 \leqslant x < 1.5$	$1.5 \leqslant x < 2.5$	$2.5 \leqslant x < 4.5$	$4.5 \leqslant x < 10$
Frequency	12	20	32	11

First calculate the midpoint of each class, xf and x^2f.

Class	Frequency (f)	Cumulative frequency	Midpoint (x)	xf	x^2f
$0 \leqslant x < 1.5$	12	12	0.75	9.0	6.75
$1.5 \leqslant x < 2.5$	20	32	2.0	40	80
$2.5 \leqslant x < 4.5$	32	64	3.5	112	392
$4.5 \leqslant x < 10$	11	75	7.25	79.75	578.188

The mean value is $\frac{\Sigma fx}{\Sigma f} = \frac{240.75}{75} = 3.21$ or 3.2 (2 s.f. as data).

The standard deviation is given by:

$$\sigma = \sqrt{\frac{\Sigma x^2}{n} - \bar{x}^2} = \sqrt{\frac{1056.938}{75} - 3.21^2} = 1.9465 \text{ or } \sigma = 2.0 \text{ (2 s.f.)}$$

To find the median and IQR, plot a cumulative frequency diagram.

Since $n = 75$ the lower quartile is the $\frac{1}{4}(75 + 1) = 19$th data item, the median is the 38th item and the upper quartile is the 57th item. From the diagram the median is 2.8, the upper quartile is 3.2 and the lower quartile is 1.9. So the IQR = upper quartile – lower quartile = 1.3.

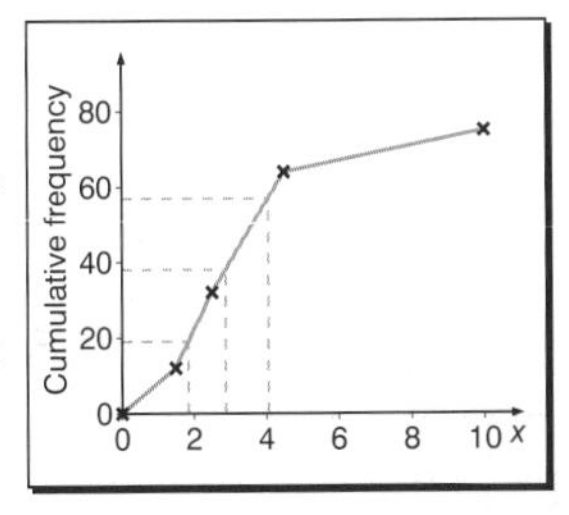

The above approach can also be used when the data are discrete. Just remember to give all final results to the nearest integer.

Symmetry in data

If the difference between the mean and the median in a set of data is zero (mean – median = 0) the distribution is **symmetrical**. The corresponding histogram or stem-and-leaf diagram is symmetric about its centre line. A distribution is **positively-skewed** (skewed to the right) if mean – median > 0; the histogram has a longer right tail than left tail. A distribution is **negatively-skewed** (skewed to the left) if mean – median < 0. The associated histogram has a longer left tail than right tail.

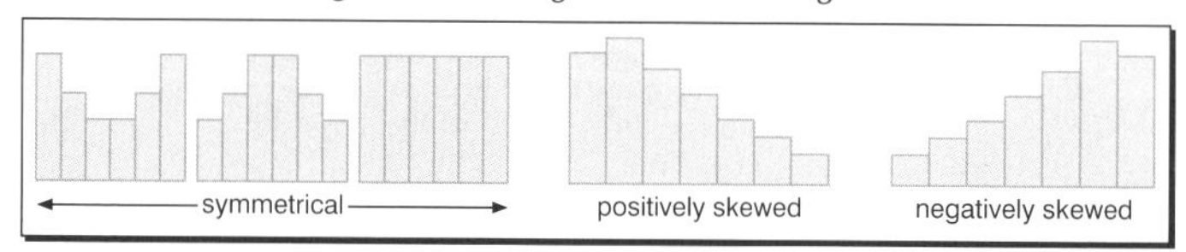

The skewness of a distribution indicates which of the measures of location and dispersion most appropriately summarises the data.

Skewed	Best measure of location	Best measure of dispersion
yes	median	IQR
no	mean	standard deviation

Box-and-whisker plot

This diagram summarises a data set in terms of its median, quartiles and range. The 'box' contains the middle half (50%) of the data (from lower to upper quartile) and the horizontal lines ('whiskers') extend to the minimum and maximum data values, giving an indication of the spread of the data. Box-and-whisker plots are a convenient method of comparing distributions.

Q Construct a box-and-whisker plot for these data.

24 27 42 96 64 81 92 63 79 29
48 41 72 43 54 5 57 35 69

Minimum value = 5, maximum value = 96. Since there are 19 data values, when they are ranked in ascending order, the lower quartile is the 5th value (35), the median is the 10th value (54) and the upper quartile is the 15th value (72). The box-and-whisker plot is drawn from these points.

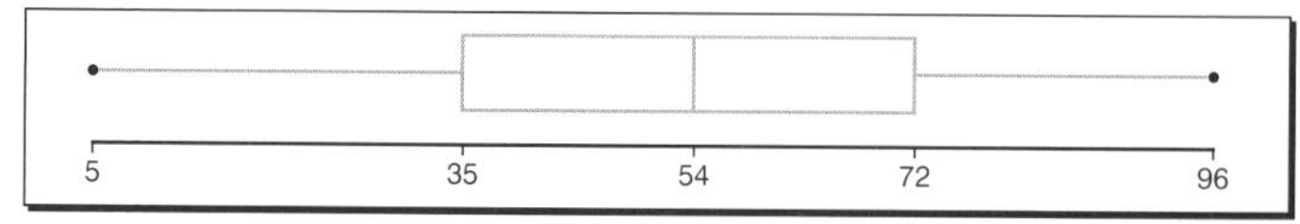

Expectation

The **expectation** or **expected value** of a function is the mean value of the function.

- In a game of chance, expected gain = E(gain) = Σgain $\times$ probability. For a fair game, E(gain) = 0.
- For a probability distribution the expected value is $E(X)$:
  $E(X) = \sum_i x_i \times P(X = x_i)$
  and the variance is given by $E(X^2) - (\text{mean})^2$.

Q In a game of shoveha'penny, players roll four 2p coins down an inclined groove onto a horizontal table on which is drawn a set of parallel lines. If the coin stops between the lines it is a 'winner' but if it stops across or is touching a line it is a 'loser'. If all four coins are 'winners' the player receives a prize of 50p (for their 8p stake), if three are winners they receive a refund of their 8p. The discrete random variable 'number of winning coins' has the following probability function.

Number of winning coins	0	1	2	3	4
Probability	0.4822	0.3858	0.1158	0.0154	0.0008

Calculate the expected income per player. Comment on the result.

To find the expected income, redefine the probability function as:

Gain to organiser	8p	8p	8p	0p	–42p
Probability	0.4822	0.3858	0.1158	0.0154	0.0008

Then the expected income per player
$= 8 \times 0.4822 + 8 \times 0.3858 + 8 \times 0.1158 + 0 \times 0.0154 - 42 \times 0.0008$
$= 7.8$p

From the organiser's viewpoint the game is quite profitable. However, the table of probabilities shows that from the players' viewpoint the great majority (98%) come away with nothing (0.4822 + 0.3858 + 0.1158).

To encourage participation, the game needs to be more attractive to the players while still offering the organisers a reasonable return. For example a small reward (say 2p or 4p) could be given for having two winning coins.

Check yourself

Handling Data

1 Construct a stem-and-leaf diagram for the following set of data. Locate the median and comment on the skewness of the distribution.

64 75 37 48 59 50 68 65 43 52
60 79 76 83 56 88 72 65 63 79

2 A student's marks for his courseworks are 56%, 62%, 65% and 59%. The pieces of coursework are weighted in the ratio 1 : 2 : 3 : 4 respectively. Calculate the student's average coursework mark.

3 Calculate the mean and variance of the following numbers.

12, 15, 18, 14, 19, 16, 14, 13, 12, 15, 17, 18, 19, 15, 14, 13, 11, 15

4 Find the mean and variance of the following frequency distribution.

Length x (cm)	150	151	152	153	154	155
Frequency f	2	3	5	4	2	1

$\Sigma f = 17$, $\Sigma fx = 2588$ and $\Sigma fx^2 = 394016$

5 Draw a histogram for the following data of cable lengths, given to the nearest metre.

Length	0–5	6–10	11–20	21–30	31–50	51–75
Number	35	42	29	45	26	11

6 Draw a percentage cumulative frequency polygon for the following data about the masses of students and from it estimate the median and interquartile range.

Mass (kg)	**Number of students**
60–62	4
63–65	9
66–68	20
69–71	13
72–74	4

The answers are on pages 119–120.

Check yourself

7 A random sample of 19 students was selected and their heights (in cm) were recorded. Draw a box-and-whisker plot to display these data.

142 181 192 129 141 172 157 105

135 127 196 163 179 148 143 154 169

8 The graphics calculator gave the following information about this data set:

4 7 7 8 5 9 14 21 5 6 8 32 14 5 20

```
1-Var Stats              1-Var Stats
↑n=15                     x̄=11
 minX=4                   Σx=165
 Q1=5                     Σx²=2691
 Med=8                    Sx=7.9102104
 Q3=14                    σx=7.64198927
 maxX=32                 ↓n=15
```

What is the best measure of location and measure of spread for these data? Justify your answer.

9 The discrete random variable X has the probability density function tabulated below.

x	1	2	3	4	5
$P(X = x)$	0.1	0.1	0.3	0.3	0.2

Determine the expectation of x.

The answers are on page 120.

If an experiment has n possible equally likely outcomes and p of them are the specific event E then the probability of E occurring, written $\mathrm{P}(E)$, is $\frac{p}{n}$.

- **Complementary events:** if $\mathrm{P}(E) = p$ then $\mathrm{P}(E \text{ does not occur}) = 1 - p$. The statement '$E$ does not occur' is the **complement of *E*** and is denoted by $\bar{E}$ or E'. Thus $\mathrm{P}(E) + \mathrm{P}(\bar{E}) = 1$.

- **Conditional probability:** the probability of an event A occurring given that the event B has already occurred. It is written as $\mathrm{P}(A|B)$ and is defined by:

$$\mathrm{P}(A|B) = \frac{\mathrm{P}(A \text{ and } B \text{ both occur})}{\mathrm{P}(B)} = \frac{\mathrm{P}(A \cap B)}{\mathrm{P}(B)}$$

- The above equation can be rearranged to give the multiplication law: $\mathrm{P}(A \cap B) = \mathrm{P}(A|B) \times \mathrm{P}(B)$

- **Independent events:** two events are independent if the occurrence of one has no influence on the occurrence of the other (e.g. two successive throws of a die are independent).

- **Inclusive events:** for two events A and B, the inclusive event is 'A or B', written as $A \cup B$. Then $\mathrm{P}(A \cup B) = \mathrm{P}(A) + \mathrm{P}(B) - \mathrm{P}(A \cap B)$
For example during the summer the days may be 'hot or wet' or 'hot and wet'.

- **Exclusive events:** two events A and B are **mutually exclusive** if they cannot both occur together, $\mathrm{P}(A \cap B) = 0$ and so, from the addition law, $\mathrm{P}(A \cup B) = \mathrm{P}(A) + \mathrm{P}(B)$.

- **Permutation:** an ordered arrangement of k objects selected from n objects; the number of different arrangements is given by ${}^n\mathrm{P}_k$ defined as:

$${}^n\mathrm{P}_k = \frac{n!}{(n-k)!}$$

- **Combination:** the number of ways of selecting k objects from n objects when the order does not matter. It is calculated from ${}^n\mathrm{C}_k$ defined as:

$${}^n\mathrm{C}_k = \frac{n!}{k!(n-k)!} = \binom{n}{k}$$

Q Two balls are drawn at random, without replacement, from a bag containing four green balls and six white balls. What is the probability that both are green?

Let B be the event 'first ball is green' and A be the event 'second ball is green'. You need to calculate $P(A \cap B)$. Now $P(B) = \frac{4}{10}$.

If the first ball is green, when the second is chosen $P(A|B) = \frac{3}{9}$.
Since $P(A \cap B) = P(A|B) \times P(B)$ then $P(A \cap B) = \frac{2}{15}$.

Q A single card is drawn from a pack of 52 playing cards. What is the probability that it is a queen or a six?

$P(A) = P(\text{queen}) = \frac{4}{52} = \frac{1}{13}$ and $P(B) = P(\text{six}) = \frac{4}{52} = \frac{1}{13}$.

Since the events 'card is a queen' and 'card is a six' are mutually exclusive, then

$P(A \cup B) = \frac{2}{13} \Rightarrow P(\text{queen or six}) = \frac{2}{13}$.

Q A communication cable consists of N links. Each link can fail, independently of the others, with probability p. The cable fails if at least one link fails. What is the probability that the cable fails?

P(at least one link fails) = 1 – P(no links fail)

P(no links fail) = $(\text{P(any given link does not fail)})^N$, since the links can fail independently.

P(cable fails) = P(at least one link fails) = $1 - (1 - p)^N$.

Q **(a)** In how many ways can a sub-committee of three people be selected from a committee of seven people?
(b) If the committee is made up of five men and two women, in how many ways can a sub-committee of two men and one woman be formed?

(a) The order of selection is not important so combinations are required. The required number is ${}^7C_3 = \frac{7!}{3!4!} = 35$.

(b) The two men can be selected from the five in ${}^5C_2 = 10$ different ways while the one woman can be selected from the two in ${}^2C_1 = 2$ different ways. The total number of ways of forming the sub-committee is $10 \times 2 = 20$.

Q How many four-letter sequences of letters can be formed from the letters of AGROUND, regardless of whether the sequences form a proper word with a meaning or not?

The order is important (a different sequencing of the same letters corresponds to a different word) so permutations are required. The number of possible arrangements is ${}^7P_4 = \frac{7!}{3!} = 840$.

Q How many permutations are there of the letters of RARER?

If all five letters were distinct then the answer would be ${}^5P_5 = \frac{5!}{0!} = 120$. However since three of the letters are identical then they can be interchanged (in 3! ways) without affecting the permutation. Thus the number of distinct permutations is $\frac{5!}{3!} = 20$.

This argument can be extended to the case where there are several distinct classes of identical objects (e,g. numbers of permutations of the letters of HAWAII is $\frac{6!}{2!2!} = 180$).

Check yourself

Probability

1 A card is drawn from a pack of 52 cards and a fair die is thrown. What is the probability that the card is a heart and the die shows a six?

2 A five-sided die numbered 1, 2, 3, 4, 5 and two coins are thrown. What is the probability of at least one tail and a multiple of 2 appearing?

3 A jar contains 50 red sweets and 40 green sweets. Two sweets are taken out (without replacement). Find the probability that:
(a) both sweets are green
(b) they are different colours.

4 A standard pack of 52 cards is shuffled and the top card is examined. The card is returned to the pack which is shuffled again. The top card is again examined. What is the probability that:
(a) both cards are red
(b) the two cards are different colours
(c) both cards are picture cards
(d) the first card is an ace and the second is black?

5 How many distinct arrangements of the letters of the following words are there?
(a) MUMMY
(b) LIZZIE
(c) TIGGER

6 How many different hands of seven cards is it possible to deal from an ordinary pack of 52 cards?

The answers are on page 121.

Bivariate data

Bivariate data are data with two variables (x, y). They are represented graphically in a scattergraph, which may suggest a link or **correlation** between the variables.

A quantitative measure of the extent to which the data are correlated is given by **Pearson's product moment correlation coefficient**, r, defined as

$$r = \frac{s_{xy}}{s_x s_y} \text{ where } s_{xy} = \frac{1}{n}\Sigma xy - \overline{x}\,\overline{y}$$

and s_x, s_y are the **standard deviations** of the x- and y-values respectively. The value of r lies in the range $-1 \leqslant r \leqslant 1$. The closer $|r|$ is to 1 the greater the degree of linear correlation in the data. If $|r|$ is close to 0 it is unlikely that there is any linear correlation in the data. The sign of r indicates whether the correlation is positive or negative.

Least squares regression

If $|r|$ is reasonably close to 1 this supports the view that there is a linear correlation between the data. When plotting the line, you must take care deciding which is the **independent** variable and which is the **dependent** (or **response**) variable. If x is the independent variable, the equation of the line of regression of y on x is:

$$\hat{y} - \overline{y} = \frac{s_{xy}}{s_x^{\,2}}(x - \overline{x})$$

If y is considered to be the independent variable, the line of repression of x on y is:

$$\hat{x} - \overline{x} = \frac{s_{xy}}{s_y^{\,2}}(y - \overline{y})$$

The symbol $\hat{}$ denotes an estimated value.

Q An electric fire was turned on in a cold room and the subsequent room temperature (y°C) was noted at 5-minute intervals after switching on (x minutes).

x (minutes)	0	5	10	15	20	25	30
y (°C)	5.4	6.5	8.4	10.5	12.7	14.7	16.7

(a) Plot a scattergraph.
(b) Explain why a line of regression $y = ax + b$ is appropriate and find its equation.
(c) Predict the room temperature after 1 hour and comment on the result.
(d) If the time to reach specified temperatures (e.g. 8°C, 10°C, 12°C, ...) had been measured, how would the calculation have been modified?

(a) The scattergraph of y (vertically) against x suggests a strong, positive linear relationship. Alternatively, determine r. Its closeness to +1 would similarly confirm the relationship.

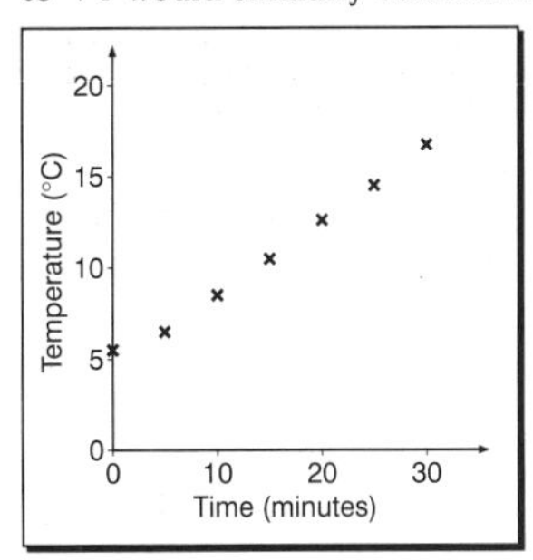

(b) Since the temperature depends on how long the fire has been switched on, y is the dependent variable.

$\bar{x} = 15$, $\bar{y} = 10.7$, $s_x = 10$, $s_{xy} = 39$ so the equation of the regression line is $\hat{y} = 0.39x + 4.85$.

(c) When $x = 60$, $\hat{y} = 28.25$. This is very warm. The model is now inappropriate as the room does not continue to warm up for ever.

(d) If the time to reach specified temperatures was measured then time (x) becomes the dependent variable and so a regression line of x on y would be required.

Normal distribution

The normal distribution curve is symmetric about the mean μ and is bell-shaped.

The standardised normal distribution has a mean of zero and a standard deviation of 1. Any variable (x) which is normally distributed with a mean μ and standard deviation σ can be converted to the standardised normal distribution (for z) using the equation $z = \frac{x - \mu}{\sigma}$.

68% of the data lie in the range $\mu \pm \sigma$, 95.5% in the range $\mu \pm 2\sigma$ and 99.7% in $\mu \pm 3\sigma$.

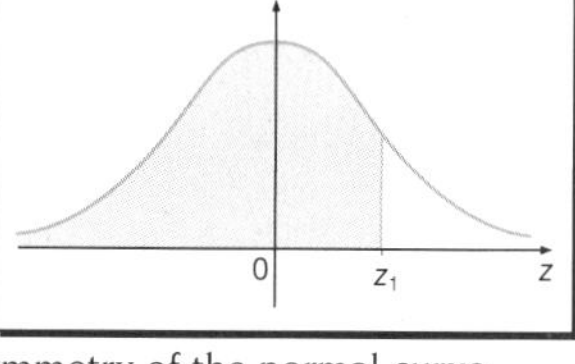

For a continuous variable, probability is represented by the area under the curve, called the **probability density function**, provided that the total area under the curve is unity (1). There are tables of the areas under the normal curve corresponding to different values of z. Areas are usually only tabulated for $z \geq 0$, other probabilities can be found from the symmetry of the normal curve. Always draw a sketch to help decide which area is required.

The shaded area represents $P(z < z_1)$.

Q The masses of sacks of compost are normally distributed with a mean of 19.6 kg and a standard deviation of 1.9 kg. Find the probability that a sack chosen at random will have a mass:

(a) greater than 22.0 kg

(b) between 19.0 and 21.0 kg.

$\mu = 19.6, \sigma = 1.9, z = \frac{x - \mu}{\sigma}$

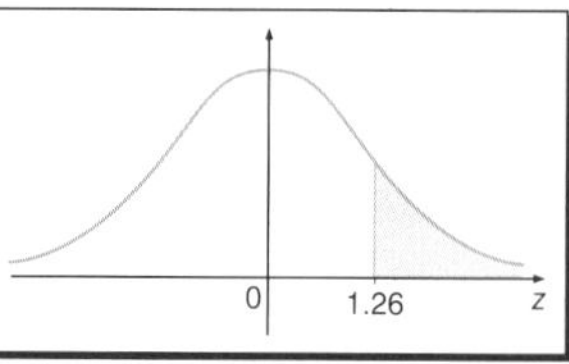

(a) Convert $x = 22.0$ to

$$z = \frac{22.0 - 19.6}{1.9} = 1.26.$$

$P(x > 22.0) \approx P(z > 1.26)$
$= 1 - 0.8962 = 0.1038$

(b) $x = 19.0 \Rightarrow z = -0.32$ and $x = 21.0 \Rightarrow z = 0.74$.

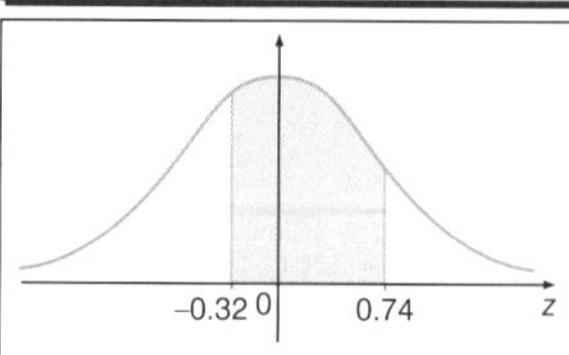

$P(19.0 < x < 21.0) \approx P(-0.32 < z < 0.74)$

Using symmetry this is equal to
$P(z < 0.74) - (1 - P(z < 0.32))$
$= 0.7704 - (1 - 0.6255) = 0.3959$

Check yourself

Bivariate Data, Normal Distribution

1 These data represent the time to exhaustion for a constant motor exercise when different quantities of glucose have been consumed before exercise.

Glucose consumed (x)	1	2	3	4	5
Time to exhaustion (y)	23	25	34	40	35

(a) Calculate the correlation coefficient.
(b) Is there strong evidence of a linear relationship between x and y?
(c) Find the equation of the line of best fit.

2 Find the probability that an observation from a standard normal distribution will be:
(a) less than 1.35
(b) less than −1.0
(c) greater than 1.85
(d) between 0.6 and 1.3.

3 The lengths of a particular component of an engine can be modelled by a normal distribution with mean 10.1 cm and standard deviation 0.5 cm. Find the probability that a randomly selected component will have a length which is:
(a) less than 10.3 cm
(b) 9.8 cm or more
(c) between 9.9 cm and 10.4 cm.
(d) Determine the lengths between which 95% of the components lie.

4 The examination marks obtained in mathematics examination are normally distributed with a mean of 55% and a standard deviation of 15%.
(a) If a grade A is awarded for marks of 70% and higher, what proportion of candidates obtain a grade A?
(b) What is the probability that a randomly chosen candidate who failed the examination had a mark less than 30%?
(c) What is the probability that a randomly chosen candidate has a mark between 30% and 39%?

The answers are on page 122.

Check yourself answers

ALGEBRA (page 11)

1 Multiply both sides by $3(2-3x)$.
$3x = 2(2-3x)$, $3x = 4-6x$, $9x = 4$, $x = \frac{4}{9}$

2 Multiply both sides by $3(2x-3)$.
$3(x+1) = 2(2x-3)$, $3x+3 = 4x-6$, $9 = x$, $x = 9$

3 The lowest common denominator is $(1+x)(1-x)$.

$$\frac{1}{1+x}\times\frac{1-x}{1-x}-\frac{1}{1-x}\times\frac{1+x}{1+x}=\frac{(1-x)-(1+x)}{(1+x)(1-x)}=\frac{-2x}{1-x^2}=\frac{2x}{x^2-1}$$

4 The lowest common denominator is $(2-x)(2+x)$.

$$\frac{1}{2-x}\times\frac{2+x}{2+x}-\frac{1}{2+x}\times\frac{2-x}{2-x}=\frac{(2+x)+(2+x)}{(2+x)(2-x)}=\frac{4}{4-x^2}$$

5 Find the lowest common denominator by factorising the denominators.
$4-x^2 = (2+x)(2-x)$ and $x^2+2x = x(2+x)$ so the lowest common denominator is $x(2+x)(2-x)$.

$$\frac{2}{(2+x)(2-x)}\times\frac{x(2+x)(2-x)}{x(2+x)(2-x)}+\frac{1}{x(2+x)}\times\frac{x(2+x)(2-x)}{x(2+x)(2-x)}$$

$$=\frac{2x}{x(2+x)(2-x)}+\frac{2-x}{x(2-x)(2-x)}=\frac{x+2}{x(2+x)(2-x)}=\frac{1}{x(2-x)}$$

6 $T^2 = 4\pi^2\frac{l}{g} \Rightarrow T^2g = 4\pi^2 l \Rightarrow \frac{T^2g}{4\pi^2} = l \Rightarrow l = \frac{T^2g}{4\pi^2}$

7 $I^2 = \frac{H}{Rt} \Rightarrow I = \sqrt{\frac{H}{Rt}}$

8 $2R = dACv^2 \Rightarrow \frac{2R}{dAv^2} = C \Rightarrow C = \frac{2R}{dAv^2}$

9 $2lf = \sqrt{\frac{T}{m}} \Rightarrow 4l^2f^2 = \frac{T}{m} \Rightarrow 4l^2f^2m = T \Rightarrow m = \frac{T}{4l^2f^2}$

10 $2x-5 > x+2 \Rightarrow 2x > x+7 \Rightarrow x > 7$

11 If $4-3x > 0$ then $4-3x \geqslant 2 \Rightarrow x < \frac{4}{3}$ and $x \leqslant \frac{2}{3} \Rightarrow x \leqslant \frac{2}{3}$

If $4-3x < 0$ then $3x-4 \geqslant 2 \Rightarrow x > \frac{4}{3}$ and $x \geqslant 2 \Rightarrow x \geqslant 2$

Combining both conditions gives $-\infty < x \leqslant \frac{2}{3}$ and $2 \leqslant x < \infty$.

12 $6-x-x^2 = (3+x)(2-x)$
The roots are 2 and –3. This is an n-shaped quadratic so the region required is the part above the x-axis i.e. $-3 < x < 2$.

13 (a) $\dfrac{2p^2q^5r^7}{(3pq^2r^3)^2} = \dfrac{2p^2q^5r^7}{9p^2q^4r^6} = \dfrac{2}{9}qr$

(b) $\sqrt[3]{2\dfrac{10}{27}} = \left(\dfrac{64}{27}\right)^{\frac{1}{3}} = \left(\dfrac{4^3}{3^3}\right)^{\frac{1}{3}} = \dfrac{4}{3}$

14 $2n^{-3} = \dfrac{1}{32} \Rightarrow n^{-3} = \dfrac{1}{64} \Rightarrow \dfrac{1}{n^3} = \dfrac{1}{64} \Rightarrow n^3 = 64 \Rightarrow n = 4$

15 $2\sqrt{6} - \sqrt{150} + \sqrt{216} = 2\sqrt{6} - \sqrt{25 \times 6} + \sqrt{36 \times 6}$
$= 2\sqrt{6} - 5\sqrt{6} + 6\sqrt{6} = 3\sqrt{6}$

16 $\dfrac{3-\sqrt{2}}{3+\sqrt{2}} \times \dfrac{3-\sqrt{2}}{3-\sqrt{2}} = \dfrac{9-6\sqrt{2}+2}{9-2} = \dfrac{11-6\sqrt{2}}{7}$

17 $2x^4 + 5x^3 - 23x^2 - 38x + 24 = (x^2 - x - 6)(ax^2 + bx + c)$
$\Rightarrow a = 2, b = 7, c = -4$
So $\dfrac{2x^4 + 5x^3 - 23x^2 - 38x + 24}{x^2 - x - 6} = 2x^2 + 7x - 4$

18 $4x^4 - 2x^3 - 11x^2 - 5x = (x^2 + x)(ax^2 + bx + c)$
$\Rightarrow a = 4, b = -6, c = -5$
So $\dfrac{4x^4 - 2x^3 - 11x^2 - 5x}{x^2 + x} = 4x^2 - 6x - 5$

19 From the second equation, $y = 2x - 3$.
Substituting this into the first equation gives:
$3x + 2(2x - 3) = 8 \Rightarrow x = 2$
Substituting this value for x into $y = 2x - 3$ gives $y = 1$.

20 From the second equation $x = 4 - 2y$.
Substituting this into the first equation gives:
$3(4 - 2y) + 2y = 6 \Rightarrow y = \frac{3}{2}$
Substituting this value for y into $x = 4 - 2y$ gives $x = 1$.

21 A sketch graph shows that there will be two solutions to this problem.

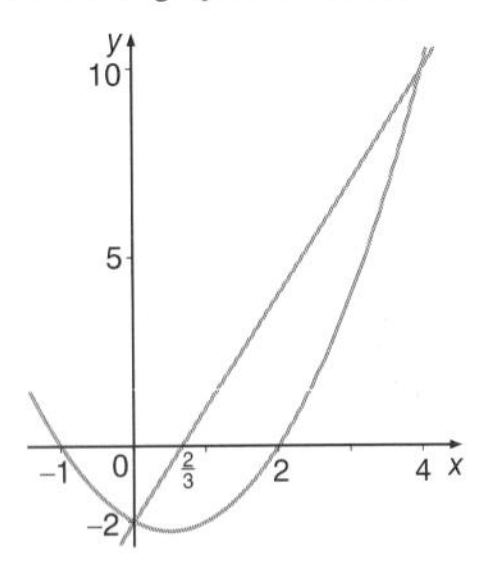

Substituting $y = 3x - 2$ into the quadratic equation:
$x^2 - 4x = 0 \Rightarrow x(x - 4) = 0 \Rightarrow x = 0$ and $x = 4$.
The corresponding values of y are -2 and 10.
So the solutions are $(0, -2)$ and $(4, 10)$.

LINEAR FUNCTIONS (page 16)

1 A is the point ((0, 4), B is the point (3, 0).
C is the point $(\frac{1}{2}(0 + 3), \frac{1}{2}(4 + 0)) = (1.5, 2)$.
Length AB $= \sqrt{3^2 + 4^2} = 5$.
Gradient of AB $= \frac{0 - 4}{3 - 0} = -\frac{4}{3}$.
The equation of AB is $y - 0 = -\frac{4}{3}(x - 3) \Rightarrow 3y + 4x = 12$.
Gradient of OC is $\frac{2 - 0}{1.5 - 0} = \frac{4}{3}$.
The equation of OC is $y - 0 = \frac{4}{3}(x - 0) \Rightarrow 3y - 4x = 0$.

2 The gradient of AB is $\frac{-2 - 4}{3 - 1} = -3$.
The equation of AB is $y - 4 = -3(x - 1) \Rightarrow y = -3x + 7$.
The gradient of the perpendicular to AB $= \frac{1}{3}$.
The equation of line through (3, −2) with gradient $\frac{1}{3}$ is
$y - -2 = \frac{1}{3}(x - 3) \Rightarrow 3y = x - 9$.

3 A graph of l against M (horizontally) shows that the points exhibit a strong linear relationship. Plot the mean point (0.661, 20.7) and draw the best straight line through the points.

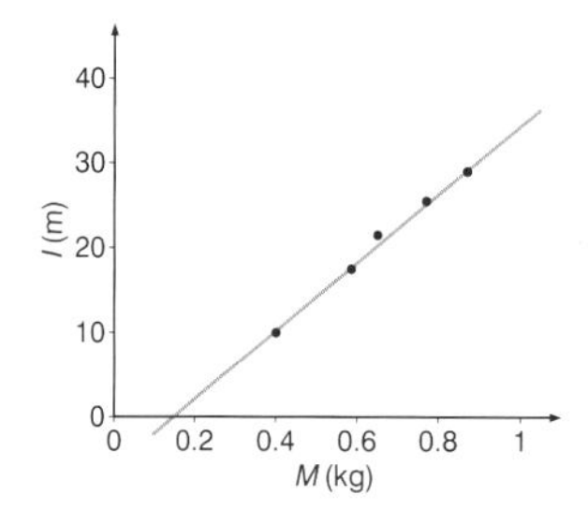

The gradient ≈ 0.0244, intercept ≈ 0.155 so the equation of the line is $l = 0.0244M + 0.155$. Unstretched length $= 0.155$ m.
When $l = 0.8$, $0.8 = 0.0244M + 0.155 \Rightarrow M = 26.4$ kg.

4 A graph of time T against age A (horizontally) shows a weak linear relationship. Plot the mean point (15.5, 13.84) and draw the best straight line through the points. Gradient ≈ –1.1, intercept ≈ 30.5. (You may not be able to read this from the graph, so calculate using the gradient and mean point.)

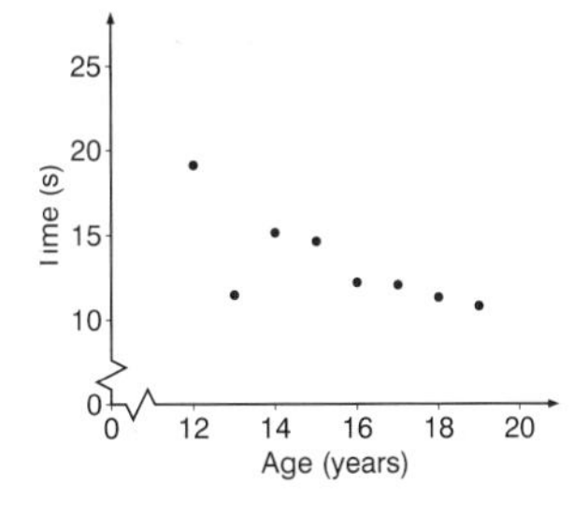

The equation of the line is $T = -1.1A + 30.5$. Using this equation gives a time of 8.5 seconds for age 20 and –2.5 seconds for age 30. Remember the warnings of using the equation to predict outside the range of the data – it is clearly absurd in this case!

Check yourself answers

QUADRATIC FUNCTIONS (page 21)

1 Completing the square on the expression $2x - 3 - x^2$ gives $-2 - (x - 1)^2$. The graph of $y = -x^2$ has the form shown.
The graph of $y = -2 - (x - 1)^2$ is obtained by moving the basic shape one unit to the right and then two units down, as shown.

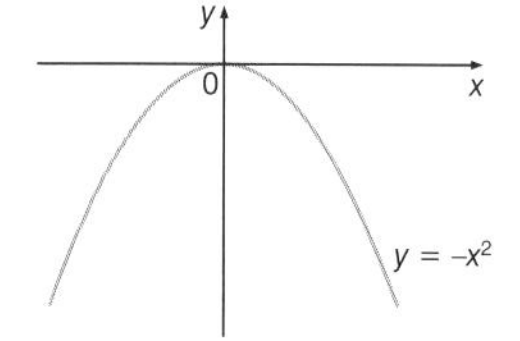

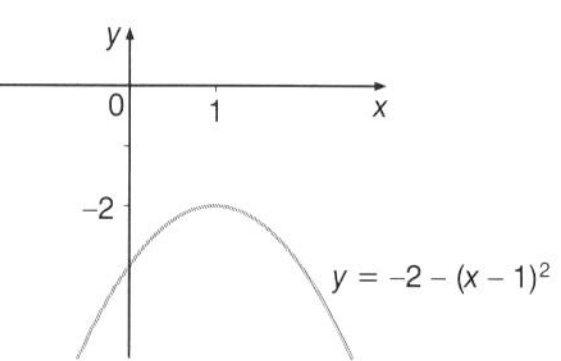

2 $(2x - 1)(x + 3) - (2x + 1)^2 = 2x^2 - x + 6x - 3 - (4x^2 + 4x + 1)$
$= -2x^2 + x - 4$

3 $2x^2 - 7x - 4$ can be written as $(2x + a)(x + b)$ where $a + 2b = -7$ and $ab = -4$.
Considering the possible values of $(a, b) \Rightarrow a = 1$ and $b = -4$.
Hence $2x^2 - 7x - 4 = (2x + 1)(x - 4)$.
The roots of $2x^2 - 7x - 4 = 0$ are $x = -\frac{1}{2}$ and $x = 4$.

4 $\dfrac{2x^2 + 7x + 6}{x^2 - 4} = \dfrac{(2x + 3)(x + 2)}{(x + 2)(x - 2)} = \dfrac{2x + 3}{x - 2}$

5 Multiplying through by x to clear the fraction gives the quadratic equation $x^2 - x - 1 = 0$.
This cannot be factorised, so use the quadratic formula.
$x = \dfrac{1 \pm \sqrt{(-1)^2 - 4 \times 1 \times -1}}{2 \times 1} = \dfrac{1 \pm \sqrt{5}}{2} = \frac{1}{2}(1 \pm \sqrt{5})$

6 $2 - x - x^2 = \frac{9}{4} - (x + \frac{1}{2})^2$
The graph is obtained by shifting the graph of $y = -x^2$ half a unit to the left and then up by $\frac{9}{4}$ units.

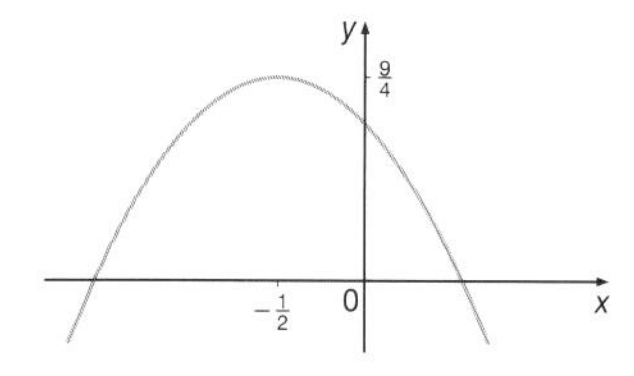

This graph intersects the x-axis twice so $2 - x - x^2 = 0$ has two roots.
Applying the 'formula' gives $x = \dfrac{1 \pm \sqrt{(-1)^2 - 4 \times -1 \times 2}}{2 \times -1} = \dfrac{1 \pm \sqrt{9}}{-2}$
$= -2$ or 1.

Check yourself answers

CUBIC AND POLYNOMIAL FUNCTIONS (page 24)

1 A quartic (polynomial of degree 4) divided by a quadratic is a polynomial of degree 2.
Let $2x^4 + 5x^3 - 23x^2 - 38x + 24 = (6 + x - x^2)(ax^2 + bx + c)$
Compare coefficients on both sides.
x^4: $a = -2$; x^3: $5 = a - b \Rightarrow b = -7$; constants: $24 = 6c \Rightarrow c = 4$
Hence $\frac{2x^4 + 5x^3 - 23x^2 - 38x + 24}{6 + x - x^2} = 4 - 7x - 2x^2$
$= (1 - 2x)(4 + x)$.

2 **(a)** Let $f(x) = x^3 - 3x^2 + 4$.
Then $f(-1) = 0$ so $x = -1$ is a root.
(b) Hence one factor of $f(x) = x^3 - 3x^2 + 4$ is $(x + 1)$.
(c) $f(x)$ can be written as a product of the linear factor and a quadratic factor:
$x^3 - 3x^2 + 4 = (x + 1)(ax^2 + bx + c)$
Comparing coefficients on both sides gives $a = 1$, $b = -4$ and $c = 4$.
$x^3 - 3x^2 + 4 = (x + 1)(x^2 - 4x + 4)$
$= (x + 1)(x - 2)^2$
The cubic has two roots: $x = -1$ and $x = 2$ but $x = 2$ is a repeated root.
(d) Consider the signs of the two factors $(x + 1)$ and $(x - 2)^2$:

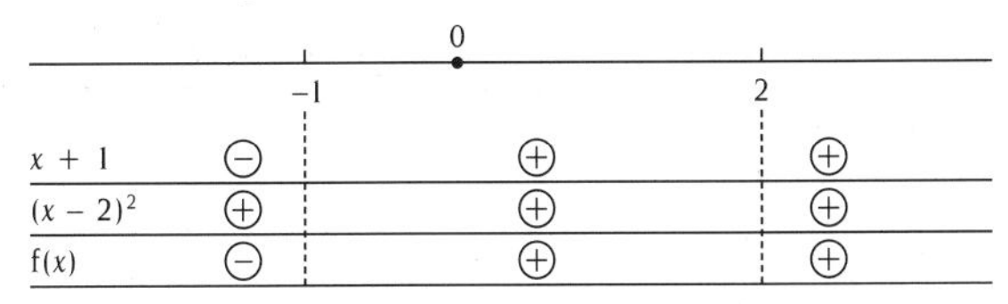

Hence $x^3 - 3x^2 + 4 > 0$ for $-1 < x < \infty$.

3 **(a)** $(x - 3)$ is a factor of $f(x)$ if $f(3) = 0$.
It is, so $(x - 3)$ is a factor.
(b) $x^3 - 7x - 6 = (x - 3)(ax^2 + bx + c)$
Comparing coefficients on both sides gives $a = 1$, $b = 3$ and $c = 2$.
So $x^3 - 7x - 6 = (x - 3)(x^2 + 3x + 2)$
(c) $f(x) = (x - 3)(x + 1)(x + 2) = 0$ so the solutions are $x = 3, -1$ and -2.
(d) $f(x + 1) = 0$ means that $(x + 1)$ is a factor so $x + 1 = 3, -1$ and $-2 \Rightarrow x = 2, -2$ and -3.

Check yourself answers

DIFFERENTIATION (page 29)

1 $f(x) = x^3(x-2)^2 = x^5 - 4x^4 + 4x^3$ so $f'(x) = 5x^4 - 16x^3 + 12x^2$
The stationary points are given by $5x^4 - 16x^3 + 12x^2 = 0$.
$x^2(5x^2 - 16x + 12) = 0$ or $x^2(5x-6)(x-2) = 0$, giving $x = 0$ (a repeated root), $x = \frac{6}{5}$ and $x = 2$. The stationary points are (0, 0), (1.2, 1.1) and (2, 0).
To classify them consider the slope, $f'(x)$, either side of each stationary point.
For (0, 0): $f'(-0.1) > 0$ and $f'(0.1) > 0$ so (0, 0) is a point of inflexion.
For (1.2, 1.1): $f'(1.1) > 0$ and $f'(1.3) < 0$ so (1.2, 1.1) is a local maximum.
For (2, 0), $f'(1.9) < 0$ and $f'(2.1) > 0$ indicating that (2, 0) is a local minimum.

2 $f(x) = x^{\frac{1}{2}} + x^{-\frac{1}{2}} \Rightarrow f'(x) = \frac{1}{2}x^{-\frac{1}{2}} - \frac{1}{2}x^{-\frac{3}{2}}$
For maximum or minimum: $\frac{1}{2}x^{-\frac{1}{2}} - \frac{1}{2}x^{-\frac{3}{2}} = 0 \Rightarrow \frac{1}{2}x^{-\frac{1}{2}}(1 - x^{-1}) = 0 \Rightarrow x = 1$
The stationary point is (1, 2). $f'(0.9) < 0$ and $f'(1.1) > 0$ so (1, 2) is a minimum.

3 The slope of the tangent to the curve is given by $f'(x) = 6x^2 + 6x - 12$.
At (–1, 15) the tangent has slope –12. The equation of the tangent at (–1, 15) is $y - 15 = -12(x - (-1)) \Rightarrow y + 12x = 3$. The slope of the normal is $+\frac{1}{12}$ so the equation of the normal at (–1, 15) is $y - 15 = \frac{1}{12}(x + 1) \Rightarrow 12y - x = 181$.

4 $f'(x) = \frac{1}{x} - 2x + 1$ and $f'(x) = 0 \Rightarrow \frac{1}{x} - 2x + 1 = 0 \Rightarrow 2x^2 - x - 1 = 0$
Solving this quadratic gives $(2x+1)(x-1) = 0$ so $x = -\frac{1}{2}$ and $x = 1$.
The solution $x = -\frac{1}{2}$ can be rejected because the logarithm function $\ln x$ is not defined for $x \leqslant 0$. The only solution is then $x = 1$ and so the stationary point is (1, 0). When $x = 0.9$, $f'(x) > 0$ and when $x = 1.1$, $f'(x) < 0$ so the maximum value of 0 occurs when $x = 1$.

5 **(a)** $\frac{dy}{dx} = 3x^2 + 10x - 4$
(i) This curve crosses the y-axis at $x = 0$. The gradient when $x = 0$ is –4.
(ii) This curve crosses the x-axis at $x = -5, -2$ and 2. The gradient when $x = -5$ is 21, when $x = -2$ it is –12 and when $x = 2$ it is 28.
(b) The curve has turning points when $3x^2 + 10x - 4 = 0$. Using the quadratic formula, $x = -3.69$ and $x = 0.36$. Then the coordinates of the turning points are (–3.69, 12.6) maximum and (0.36, –20.75) minimum.

6 $2\sqrt{x}(5 - x) = 10\sqrt{x} - 2x\sqrt{x} = 10x^{\frac{1}{2}} - 2x^{\frac{3}{2}}$ and $\frac{dy}{dx} = 5x^{-\frac{1}{2}} - 3x^{\frac{1}{2}} = \frac{5 - 3x}{\sqrt{x}}$
For a stationary point $\frac{dy}{dx} = 0$ so $5 - 3x = 0 \Rightarrow x = \frac{5}{3}$ with y-value 8.61 to 3 s.f.
Investigating the gradient either side of this point shows that it is a maximum.

7 Surface area of tank $= \pi r^2 + 2\pi rh$. Volume $= \pi r^2 h = 5$. Substituting for h in the formula for the surface area gives $A = \pi r^2 + \frac{10}{r}$. Differentiating gives $\frac{dA}{dr} = 2\pi r - \frac{10}{r^2}$. Solving $\frac{dA}{dr} = 0$ gives $r = 1.168$ (3 d.p.). The corresponding value of h is 1.168. The dimensions of the tank are radius = height = 1.168 m.

8 $f(x + h) = (x + h)^3 = x^3 + 3x^2h + 3xh^2 + h^3$
$f(x + h) - f(x) = 3x^2h + 3xh^2 + h^3$
$\frac{f(x+h) - f(x)}{h} = 3x^2 + 3xh + h^2$ so taking the limit as $h \to 0$ gives $f'(x) = 3x^2$.

Check yourself answers

INTEGRATION (page 34)

1 **(a)** $f'(x) = x^2(x + 1) \Rightarrow f(x) = \frac{x^4}{4} + \frac{x^3}{3} + c$

When $x = 1$, $y = 1$ giving $c = \frac{5}{12}$ so $f(x) = \frac{x^4}{4} + \frac{x^3}{3} + \frac{5}{12}$

(b) $f'(x) = (x^2 - 1)(3x + 5) \Rightarrow f(x) = \frac{3}{4}x^4 + \frac{5}{3}x^3 - \frac{3}{2}x^2 - 5x - 3$

(c) $f'(x) = 6e^{3x} \Rightarrow f(x) = 2e^{3x} + 1$

2 **(a)** $\int_0^4 3\sqrt{x}(1 + x)dx = 3\int_0^4 (x^{\frac{1}{2}} + x^{\frac{3}{2}})dx$

$$= 3\left[\frac{2x^{\frac{5}{2}}}{5} + \frac{2x^{\frac{3}{2}}}{3}\right]_0^4 = 3 \times \frac{272}{15} = \frac{272}{5}$$

(b) $\int_2^3 \frac{x^4 - 4}{x^3}dx = \int_2^3 (x - 4x^{-3})dx = \left[\frac{x^2}{2} + \frac{2}{x^2}\right]_2^3 = \frac{20}{9}$

(c) $\int_{-2}^{-1} \frac{(x + 2)^2}{x^4}dx = \int_{-2}^{-1}\left(\frac{1}{x^2} + \frac{4}{x^3} + \frac{4}{x^4}\right)dx = \left[-\frac{1}{x} - \frac{2}{x^2} - \frac{4}{3x^3}\right]_{-2}^{-1} = \frac{1}{6}$

(d) $\int_1^3 (10x - e^x)dx = \left[5x^2 - e^x\right]_1^3 = -e^3 + e + 40$

3 **(a)** Area required $= \int_0^1 (1 - x^2 + x^4)dx = \left[x - \frac{x^3}{3} + \frac{x^5}{5}\right]_0^1 = \frac{13}{15}$

(b) Area required $= \int_{-1}^1 2e^{-x}dx = \left[-2e^{-x}\right]_{-1}^1 = -2(e^{-1} - e^1) = 2(e - e^{-1})$

4 To find the area you need to do the integral in two parts.

Area 1 $= \int_{-2}^1 (x^3 - x^2 - 4x + 4)dx = \left[\frac{x^4}{4} - \frac{x^3}{3} - 2x^2 + 4x\right]_{-2}^1 = \frac{45}{4}$

Area 2 $= \left|\int_1^2 (x^3 - x^2 - 4x + 4)dx\right| = \left|\left[\frac{x^4}{4} - \frac{x^3}{3} - 2x^2 + 4x\right]_1^2\right| = \left|\frac{-7}{12}\right| = \frac{7}{12}$

So the total area required $= \frac{7}{12} + \frac{45}{4} = \frac{71}{6}$

5 The two graphs intersect when

$\frac{3}{x} = 4 - x \Rightarrow x^2 - 4x + 3 = 0 \Rightarrow (x - 1)(x - 3) = 0 \Rightarrow x = 1, x = 3$

When $x = 1$, $y = 3$ and when $x = 3$, $y = 1$.

The volume of revolution about the y-axis $= \pi \int_1^3 x^2 dy$

The area is between $x = \frac{3}{y}$ and $x = 4 - y$.

The volume required $= \pi \int_1^3 (4 - y)^2 dy - \pi \int_1^3 (\frac{3}{y})^2 dy$

$$= \pi \int_1^3 (16 - 8y + y^2 - \frac{9}{y^2}) dy$$

$$= \pi \left[16y - 4y^2 + \frac{y^3}{3} + \frac{9}{y} \right]_1^3$$

$$= \frac{8\pi}{3}$$

6 The two graphs intersect when $x(4 - x) = x$

$4x - x^2 = x \Rightarrow x^2 - 3x = 0 \Rightarrow x(x - 3) = 0 \Rightarrow x = 0, x = 3$

Thus the point of intersection is the point (3, 3).

The volume generated when the shaded region is rotated about the x-axis

$= \pi \int_0^3 (x(4 - x))^2 dx - \pi \int_0^3 x^2 dx = \pi \int_0^3 (x^4 - 8x^3 + 16x^2) dx - \pi \int_0^3 x^2 dx$

$= \pi \left[\frac{x^5}{5} - 8\frac{x^4}{4} + 16\frac{x^3}{3} \right]_0^3 - \pi \left[\frac{x^3}{3} \right]_0^3 = \frac{108\pi}{5}$

Check yourself answers

TRIGONOMETRY (page 41)

1 $2\cos^2 x + 3\sin x = 0 \Rightarrow 2(1 - \sin^2 x) + 3\sin x = 0 \Rightarrow 2\sin^2 x - 3\sin x - 2 = 0$
Factorising the quadratic gives:
$(2\sin x + 1)(\sin x - 2) = 0 \Rightarrow \sin x = -\frac{1}{2}$ or 2
$\sin x = -\frac{1}{2} \Rightarrow x = -30°$ (or 330°) from the calculator and $180° + 30° = 210°$, as the sine function is also negative in the third quadrant.
The equation $\sin x = 2$ has no solutions (since $-1 \leqslant \sin x \leqslant 1$).
$x = 210°, 330°$

2 $F(\cos\alpha + \mu\sin\alpha) = \mu W$
Substitute $\mu = \tan\lambda$ then $F(\cos\alpha + \tan\lambda\sin\alpha) = W\tan\lambda$
Substitute $\frac{\sin\lambda}{\cos\lambda}$ for $\tan\lambda$ to give $F(\cos\alpha + \frac{\sin\lambda\sin\alpha}{\cos\lambda}) = \frac{W\sin\lambda}{\cos\lambda}$
which leads to $F(\cos\alpha\cos\lambda + \sin\alpha\sin\lambda) = W\sin\lambda$
Hence $F\cos(\alpha - \lambda) = W\sin\lambda \Rightarrow F = \frac{W\sin\lambda}{\cos(\alpha - \lambda)}$.

3
$$\begin{aligned}\sin 3\theta &= \sin(2\theta + \theta) = \sin 2\theta\cos\theta + \sin\theta\cos 2\theta \\ &= 2\sin\theta\cos^2\theta + \sin\theta(1 - 2\sin^2\theta) \\ &= 2\sin\theta(1 - \sin^2\theta) + \sin\theta(1 - 2\sin^2\theta) \\ &= 3\sin\theta - 4\sin^3\theta\end{aligned}$$
Substituting into the equation $\sin 3\theta - 2\sin\theta = 0$ gives
$3\sin\theta - 4\sin^3\theta - 2\sin\theta = 0$
$\sin\theta - 4\sin^3\theta = 0$
$\sin\theta(1 - 4\sin^2\theta) = 0 \Rightarrow \sin\theta = 0$ or $\sin^2\theta = \frac{1}{4} \Rightarrow \sin\theta = -\frac{1}{2}, \frac{1}{2}$.
Now $\sin\theta = 0 \Rightarrow \theta = 0, 180°, 360°$
$\sin\theta = -\frac{1}{2} \Rightarrow \theta = 210°, 330°$

$\sin\theta = \frac{1}{2} \Rightarrow \theta = 30°, 150°$

Hence $\theta = 0, 30°, 150°, 180°, 210°, 330°, 360°$.

4
$$\frac{2\sin\theta + \sin 2\theta}{1 - \cos 2\theta} = \frac{2\sin\theta + 2\sin\theta\cos\theta}{1 - (\cos^2\theta - \sin^2\theta)} = \frac{2\sin\theta(1 + \cos\theta)}{1 - \cos^2\theta + \sin^2\theta} = \frac{2\sin\theta(1 + \cos\theta)}{2(1 - \cos^2\theta)}$$
$$= \frac{\sin\theta(1 + \cos\theta)}{(1 - \cos\theta)(1 + \cos\theta)} = \frac{\sin\theta}{1 - \cos\theta}$$

5 Consider the expression under the cube root sign.
$$\frac{\sin\theta}{1 + \cot^2\theta} = \frac{\sin\theta}{1 + \frac{\cos^2\theta}{\sin^2\theta}} = \frac{\sin\theta\sin^2\theta}{\sin^2\theta + \cos^2\theta} = \sin^3\theta$$
So $\sqrt[3]{\frac{\sin\theta}{1 + \cos^2\theta}} = \sin\theta$

6 Using radian measure:
arc length $= r\theta = 12$ **(i)**
sector area $= \frac{1}{2}r^2\theta = 54$ **(ii)**
Divide **(ii)** by **(i)** to give $r = 9$ cm.
Substitute for r into **(i)** to give $\theta = \frac{12}{9} = 1.33$ radians.

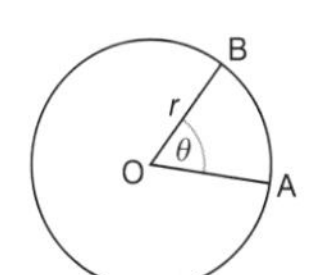

7 $h = 5 + 3\sin\frac{\pi t}{6}$

High tide ($h = 8$ m) occurs when $\sin\frac{\pi t}{6} = 1$

$\Rightarrow \frac{\pi t}{6} = \frac{\pi}{2}, \frac{5\pi}{2}, \ldots \Rightarrow t = 3, 15$

High tide occurs at 0300 hours and 1500 hours.

Low tide ($h = 5 - 3 = 2$ m) occurs when $\sin\frac{\pi t}{6} = -1$

$\Rightarrow \frac{\pi t}{6} = \frac{3\pi}{2}, \frac{7\pi}{2}, \ldots \Rightarrow t = 9, 21$

Low tide occurs at 0900 hours and 2100 hours.

Tide height of 6 m $\Rightarrow 3\sin\frac{\pi t}{6} = 1 \Rightarrow t = \frac{6}{\pi}\sin^{-1}\frac{1}{3} = 0.65$ (from calculator).

From the sketch, the graph is rising at this time (positive slope) so the time is approximately 0040 hours.

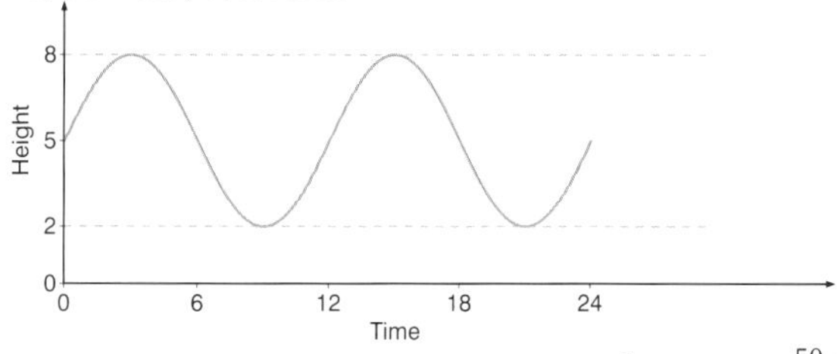

8 Arc length $= r\theta$, perimeter $= 2r + r\theta = 50$, area $= \frac{1}{2}r^2\theta$. Now $\theta = \frac{50 - 2r}{r}$ from the perimeter and so the area is $\frac{1}{2}r^2(\frac{50 - 2r}{r}) = 25r - r^2$. Differentiating to find a maximum gives $\frac{dA}{dr} = 25 - 2r$, solving $\frac{dA}{dr} = 0$ gives $r = 12.5$. For $r = 12.5$, $\theta = 2$. So the largest sector area is when the radius is 12.5 cm and the angle is 2 radians.

9 $5\cos x - 12\sin x$ is equivalent to $R\cos(x + \alpha) = R\cos\alpha\cos x - R\sin\alpha\sin x$. So $5 = r\cos\alpha$ and $12 = r\sin\alpha$, giving $R = \sqrt{5^2 + 12^2} = 13$ and $\alpha = \tan^{-1}-\frac{5}{12}$ $= -22.6°$ (from the calculator). However, since $\sin\alpha$ and $\cos\alpha$ are positive, the required value for α is 67.4°. Hence $5\cos x - 12\sin x = 13\cos(x + 67.4°)$. The maximum value of $f(x)$ is 13 and the minimum value is -13.

10 **(a)** $\text{cosec}\theta = 4 \Rightarrow \frac{1}{\sin\theta} = 4 \Rightarrow \sin\theta = \frac{1}{4}$. $\theta = \sin^{-1}0.25 = 14.47°$.
Also $180° - 14.47° = 165.53°$.

(b) $\sec 2\theta = 5 \Rightarrow \frac{1}{\cos 2\theta} = 5 \Rightarrow \cos 2\theta = \frac{1}{5}$. $2\theta = \cos^{-1}0.2 = 78.46°$
so $\theta = 39.23°$. Also $180° - 39.23° = 140.77°$, $180° + 39.23° = 119.23°$ and $360° - 39.23° = 321.77°$.

Check yourself answers

SEQUENCES AND SERIES (page 45)

1 $S_n = 2n^2 + 3n$
$S_n = S_{n-1} + u_n$ by definition since
$S_n = u_1 + u_2 + u_3 + \ldots + u_{n-1} + u_n$
$u_n = S_n - S_{n-1} = 2n^2 + 3n - \{2(n-1)^2 + 3(n-1)\} = 4n - 1$
So $u_n = 4n + 1$
For an AP $u_1 = a = 4 \times 1 + 1 \therefore a = 5$
$u_2 = a + d = 4 \times 2 + 1 \Rightarrow d = 4$

2 **(a)** $u_5 = ar^4 = 1.08$; $u_2 = ar = 40$ so $\dfrac{ar^4}{ar} = r^3 = \frac{1.08}{40} = 0.027 \Rightarrow r = 0.3$

(b) Since $r = 0.3$ and $ar = 40$ then the first term, $a = \frac{400}{3}$

(c) Since $|r| < 1$ then $S_\infty = \dfrac{a}{1-r} = \dfrac{400/3}{1-0.3} = \dfrac{4000}{21}$

(d) The sum of the first n terms $= \dfrac{400[1-(0.3)^n]}{3 \times 0.7} = \frac{4000}{21}(1 - 0.3^n)$

So $S_n - S_\infty = \frac{4000}{21}(1 - 0.3^n) - \dfrac{4000}{21}$

Thus $|S_n - S_\infty| < 0.001$ provided $(0.3)^n < \frac{0.021}{4000} = 0.000\,005$

Use trial and error for values of n to find $n = 11$ satisfies the inequality.

3 $(1 + 3x)^{-2} = 1 + (-2)(3x) + \dfrac{(-2)(-3)}{2!}(3x)^2 + \dfrac{(-2)(-3)(-4)}{3!}(3x)^3 + \ldots$
$= 1 - 6x + 27x^2 - 108x^4 + \ldots$
The series converges provided that $-1 < 3x < 1 \Rightarrow -\frac{1}{3} < x < \frac{1}{3}$ or $|x| < \frac{1}{3}$
Comparing $\dfrac{1}{(1+3x)^2}$ with $\dfrac{1}{(1.03)^2}$, $x = 0.01$.
Substitute $x = 0.01$ in the above approximation to give:
$\dfrac{1}{1.03^2} \approx 1 - 6 \times 0.01 + 27 \times 0.01^2 - 108 \times 0.01^3$
$= 1 - 0.06 + 0.0027 - 0.000\,108$

So any further terms will be negligible to 3 s.f.
Hence $\dfrac{1}{1.03^2} \approx 0.942\,592 = 0.943$ to 3 s.f.

4 Expanding to the given degree of approximation:

$$(1-4x)^{\frac{1}{2}} = 1 + \tfrac{1}{2}(-4x) + \frac{(\frac{1}{2})(-\frac{1}{2})}{2!}(-4x)^2 = 1 - 2x - 2x^2$$

$$(1+3x)^{\frac{1}{3}} = 1 + \tfrac{1}{3}(3x) + \frac{(\frac{1}{3})(-\frac{2}{3})}{2!}(3x)^2 = 1 + x - x^2$$

$$(1+x)^{-\frac{1}{2}} = 1 - \tfrac{1}{2}x + \tfrac{3}{8}x^2$$

Thus:

$$\frac{(1-4x)^{\frac{1}{2}}(1+3x)^{\frac{1}{3}}}{(1+x)^{\frac{1}{2}}} = (1-4x)^{\frac{1}{2}}(1+3x)^{\frac{1}{3}}(1+x)^{-\frac{1}{2}}$$

$$= (1-2x-2x^2)(1+x-x^2)(1-\tfrac{1}{2}x+\tfrac{3}{8}x^2)$$

$$= (1-2x-2x^2)(1+\tfrac{1}{2}x-\tfrac{9}{8}x^2) \text{ (neglecting } x^3 \text{ and higher powers)}$$

$$= (1-2x-2x^2)(1) + (1-2x-2x^2)\tfrac{1}{2}x - (1-2x-2x^2)\tfrac{9}{8}x^2$$

$$= 1 - 2x + \tfrac{1}{2}x - x^2 - \tfrac{9}{8}x^2 \text{ (neglecting } x^3 \text{ and higher powers)}$$

$$= 1 - \tfrac{3}{2}x - \tfrac{33}{8}x^2$$

5 **(a)** The series is geometric.
$a = 100$, $r = 0.9$ so $u_{20} = 100 \times 0.9^{19} = 13.5085$ to 4 d.p.
$S_{20} = 100\frac{(1-0.9^{20})}{1-0.9} = 878.4233$ to 4 d.p.

(b) The series is arithmetic.
$a = 5$, $d = 7$ so $u_{20} = 5 + (21-1) \times 7 = 138$
$S_{20} = \frac{1}{2} \times 20(2 \times 5 + (20-1) \times 7 = 1430$

6 $(1+2x)^{-2} = 1 + (-2)(2x) + \frac{(-2)(-3)}{2!}(2x)^2 + \frac{(-2)(-3)(-4)}{3!}(2x)^3 + \ldots$
$= 1 - 4x + 12x^2 - 32x^3 + \ldots$
The series will converge for $-1 \leqslant 2x \leqslant 1$ i.e. $-\frac{1}{2} \leqslant x \leqslant \frac{1}{2}$.

1 (a) $f:x \to 3x + 1 \Rightarrow f(x) = 3x + 1$, $g:x \to x^2 - 2 \Rightarrow g(x) = x^2 - 2$
So $fg(x) = f(g(x)) = 3(g(x)) + 1 = 3(x^2 - 2) + 1 = 3x^2 - 5$

(b) A sketch of $y = 3x^2 - 5$ reveals that the range of fg is $[-5, \infty]$.

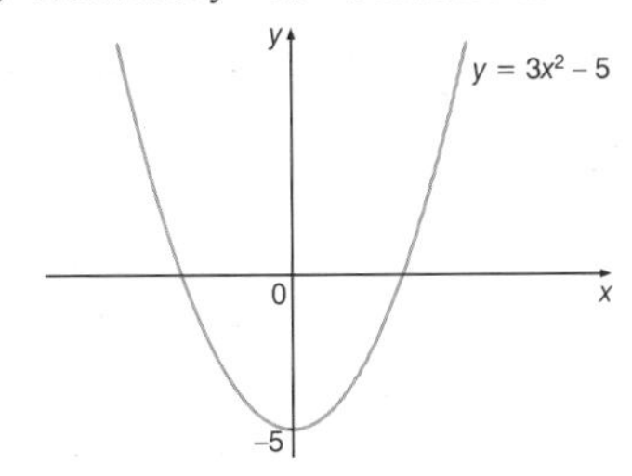

(c) To find the inverse of f, let $y = 3x + 1$.
Then $x = \frac{1}{3}(y - 1)$ so $f^{-1}(x) = \frac{1}{3}(x - 1)$

The equation $f^{-1}(x) = g(x) \Rightarrow \frac{1}{3}(x - 1) = x^2 - 2$

$\Rightarrow 3x^2 - x - 5 = 0$
Solving this quadratic (using the formula) gives
$x = \frac{1}{6}(1 \pm \sqrt{61}) = -1.14, 1.47$

2 (a) $f(x) = \frac{1}{3}x^3$ and $g(x) = x - 2$ so $f(g(x)) = \frac{1}{3}(x - 2)^3$.

(b) Let $y = \frac{1}{3}x^3$, then $x^3 = 3y$ so $x = \sqrt[3]{3y}$, hence $f^{-1}(x) = \sqrt[3]{3x}$.
Let $y = x - 2$, then $x = 2 + y$ so $g^{-1}(x) = 2 + x$.

(c) First, determine the inverse $(fg)^{-1}$ of the composite function $fg(x) = \frac{1}{3}(x - 2)^3$.
Let $y = \frac{1}{3}(x - 2)^3$, then $3y = (x - 2)^3 \Rightarrow x = 2 + \sqrt[3]{3y}$

So $(fg)^{-1}(x) = 2 + \sqrt[3]{3x}$.

Now determine the composite function.

$g^{-1}f^{-1} = g^{-1}(f^{-1}(x)) = 2 + f^{-1}(x) = 2 + \sqrt[3]{3x}$

So $(fg)^{-1} = g^{-1}f^{-1}$.

3 $f(x) = \frac{3}{x^2}$ and $f(-x) = \frac{3}{(-x)^2} = \frac{3}{x^2}$

So f(x) is even and its graph is symmetric with respect to the y-axis.

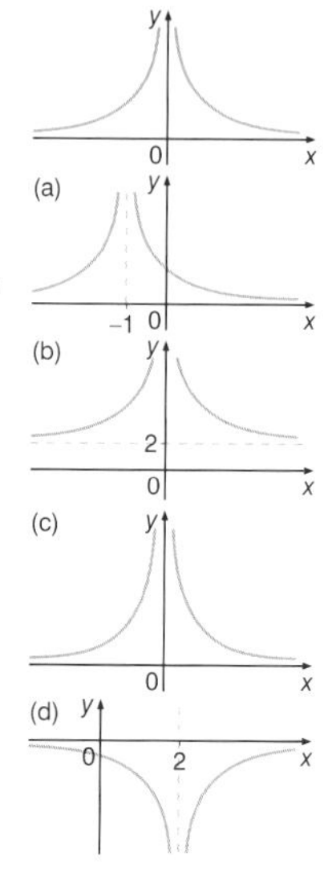

(a) A translation of −1 unit parallel to the x-axis means x is replaced by $x - (-1) = (x + 1)$

So $f_1(x) = \frac{3}{(x+1)^2}$ with asymptotes $x = -1$ and $y = 0$.

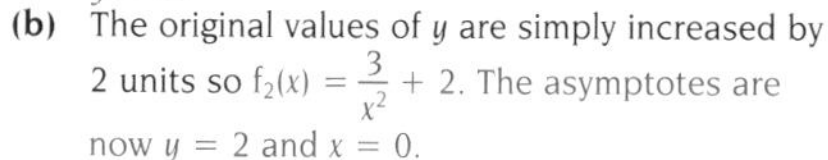

(b) The original values of y are simply increased by 2 units so $f_2(x) = \frac{3}{x^2} + 2$. The asymptotes are now $y = 2$ and $x = 0$.

(c) The original values of y are simply doubled.

$f_3(x) = 2 \times \frac{3}{x^2} = \frac{6}{x^2}$. The asymptotes are $x = 0$ and $y = 0$.

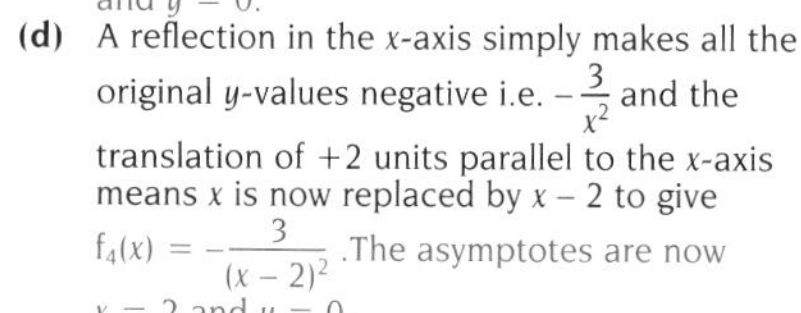

(d) A reflection in the x-axis simply makes all the original y-values negative i.e. $-\frac{3}{x^2}$ and the translation of +2 units parallel to the x-axis means x is now replaced by $x - 2$ to give

$f_4(x) = -\frac{3}{(x-2)^2}$. The asymptotes are now $x = 2$ and $y = 0$.

4 $f(x) = \pi - x$, $0 \leqslant x \leqslant \pi$ and $f(x) = -\sin x$, $\pi \leqslant x \leqslant 2\pi$

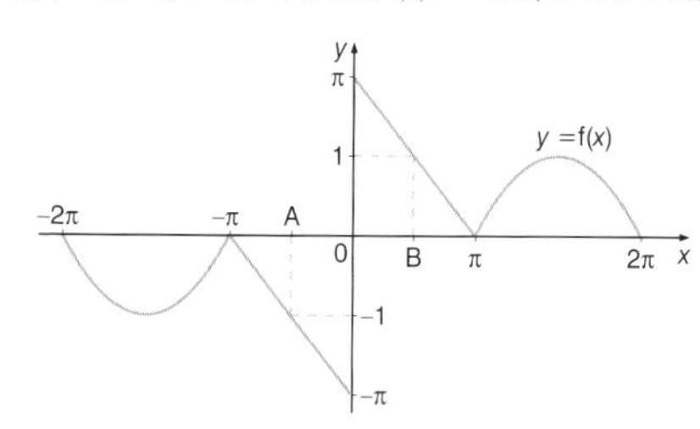

The inverse of f is only defined when the value of x corresponding to a given value of y is unique.

This is seen only to be true in this case for y-values in the range $(1, \pi]$ and $[-\pi, -1)$, corresponding to x-values in the interval [A, B].

B is found by solution of the equation $1 = \pi - x \Rightarrow x = \pi - 1$.

By symmetry A is the point $x = 1 - \pi$.

So the inverse only exists for $1 - \pi < x < \pi - 1$.

Check yourself answers

EXPONENTIALS AND LOGARITHMS (page 53)

1 (a) At $t = 0$, $C = 10 = C_0e^0$, but $e^0 = 1$ so $C_0 = 10$ and hence $C = 10e^{-0.15t}$.

(b) The graph is characteristic of the negative exponential and passes through the points (0, 10) and $(6, 10e^{-0.15 \times 6}) = (6, 4.07)$. Hence the graph is as shown.

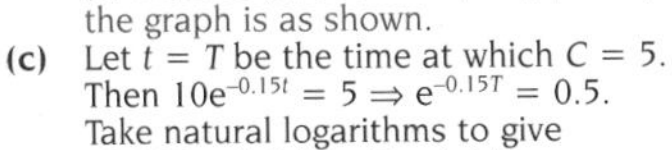

(c) Let $t = T$ be the time at which $C = 5$.
Then $10e^{-0.15t} = 5 \Rightarrow e^{-0.15T} = 0.5$.
Take natural logarithms to give
$-0.15T = \ln 0.5 \Rightarrow T \approx 4.6$ hours.

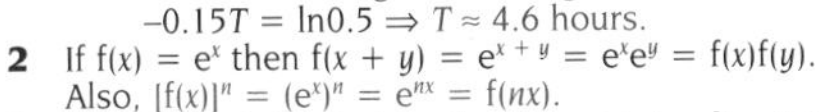

2 If $f(x) = e^x$ then $f(x + y) = e^{x+y} = e^xe^y = f(x)f(y)$.
Also, $[f(x)]^n = (e^x)^n = e^{nx} = f(nx)$.

3 (a) Multiply both sides by e^x to obtain $e^{2x} - 4e^x - 1 = 0$.
Using the laws of indices, you can write this as a quadratic.
$(e^x)^2 - 4e^x - 1 = 0$
Solve, using the formula, to give $e^x = 2 \pm \sqrt{5}$.
The only acceptable solution is $e^x = 2 + \sqrt{5}$ (since the exponential function is never negative).
So $e^x = 2 + \sqrt{5} \Rightarrow x = \ln(2 + \sqrt{5}) = 1.44$

(b) $\ln(x + 1) - \ln x = 0.1 \Rightarrow \ln\left(\frac{x+1}{x}\right) = 0.1$ (using the properties of logarithms).

$$\frac{x+1}{x} = e^{0.1} \Rightarrow x + 1 = xe^{0.1} \Rightarrow 1 = x(e^{0.1} - 1) \Rightarrow x = \frac{1}{e^{0.1} - 1} \approx 9.51.$$

4 $y = \ln 3x = \ln 3 + \ln x$ using the properties of logarithms.
The graph of this function can be obtained from that of $y = \ln x$ by shifting it vertically upwards by an amount $\ln 3$.

5 To test for an exponential relationship of the form $y = ae^{bx}$ plot a graph of $\ln y$ vertically against x.

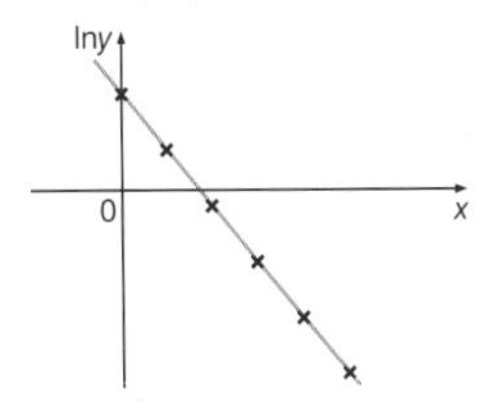

The points give a good straight line, which means that the data obey a law of the form $y = ae^{bx}$. The intercept on the vertical axis $= 0.69 \Rightarrow a = e^{0.69} \approx 2$ and

$$b = \text{slope of line} = \frac{-1.830 - 0.69}{25 - 0} = \frac{-2.52}{25} = -0.1.$$

Thus the data obey the relationship $y = 2e^{-0.1x}$.

Check yourself answers

NUMERICAL METHODS (page 57)

1 The equation can be written as $f(x) = x^2 - x - 1 = 0$ so $f(1.5) = 0.25$ and $f(2) = +1$. The sign change in the two function values indicates a root (solution) in $[1.5, 2]$. The bisection method calculations can be set out as follows.

a	f(a)	b	$f(b)$	$c = \frac{1}{2}(a + b)$	$f(c)$	Interval width
1.5	−0.25	2	1	1.75	0.3125	0.5
1.5	−0.25	1.75	0.3125	1.625	0.0156	0.25
1.5	−0.25	1.625	0.0156	1.5625	−0.1211	0.125
1.5625	−0.1211	1.625	0.0156	1.59375	0.0537	0.0625

Since the interval width is now less than 0.1, the midpoint $x = 1.593\,75$ can be taken as the approximation to the root, as its maximum error is less than $\frac{0.1}{2} = 0.05$ as required.

2 $x^2 - x - 1 = 0 \Rightarrow x^2 = 1 + x \Rightarrow x = \frac{1 + x}{x} = 1 + \frac{1}{x}$ so $g(x) = 1 + \frac{1}{x}$.

Thus $g'(x) = -\frac{1}{x^2}$ and $|g'(x_0)| = \frac{1}{1.5^2} < 1$ so the iterations can be expected to converge to a root. The iterations obtained by the graphics calculator are shown, beginning with $x_0 = 1.5$. To 3 decimal places, the root = 1.618.

```
1.5
              1.5
1+1/Ans
      1.666666667
              1.6
            1.625
      1.615384615
■
```

```
1.615384615
1.619047619
1.617647059
1.618181818
1.617977528
1.618055556
1.618025751
```

3 Let $f(x) = x^2 - x - 1$, then $f'(x) = 2x - 1$.

Newton Raphson gives $x_{n+1} = x_n - \frac{x_n^2 - x_n - 1}{2x_n - 1}$.

The iterations are as shown, starting with $x_0 = 2$.

```
2
                 2
Ans-(Ans²-Ans-1)
/(2Ans-1)
       1.666666667
       1.619047619
       1.618034448
■
```

Thus after four iterates the root can be confirmed as 1.618, to 3 d.p.

4 Inspecting the x-values shows that $h = 2$. Inspecting the y-values shows $y_0 = 0$, $y_1 = 12$, ... , $y_5 = 300$. Hence the trapezium rule gives:

$$\int_0^{10} y\,dx \approx \tfrac{2}{2}\{(0 + 300) + 2(12 + 48 + 108 + 192)\} = 1020$$

Check yourself answers

FORCES (page 63)

1 The resultant force $\mathbf{R}_b$ on the box is given by:
$\mathbf{R}_b = T\mathbf{i} + F\mathbf{i} + N\mathbf{j} - 60g\sin30^\circ\mathbf{i} - 60g\cos30^\circ\mathbf{j}$.
Since the box is on the point of sliding, $F = 0.4R$ where R is the normal reaction force. So $\mathbf{R}_b = (T + 0.4R - 60g\sin30^\circ)\mathbf{i} + (R - 60g\cos30^\circ)\mathbf{j}$. Since the box is in equilibrium, $\mathbf{R}_b = 0$ and so
$T + 0.4R - 60g\sin30^\circ = 0$ and
$R - 60g\cos30^\circ = 0$.
Solving for T gives $T = 90.3\,\text{N}$.

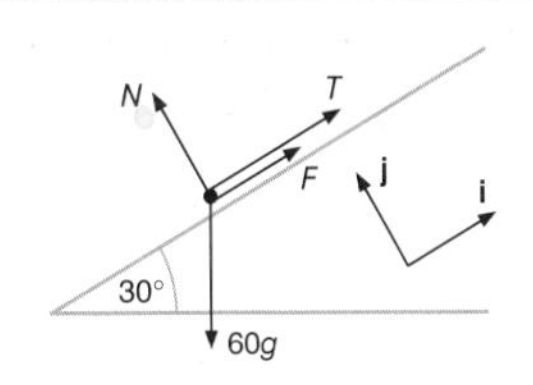

2 **(a)** Pulled up the slope

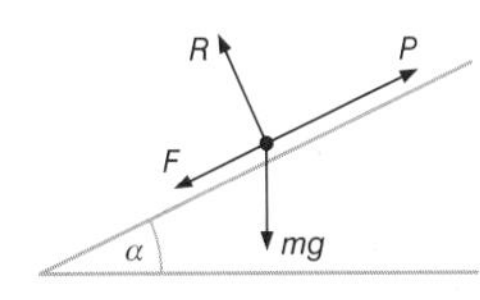

(b) Sliding down the slope under gravity

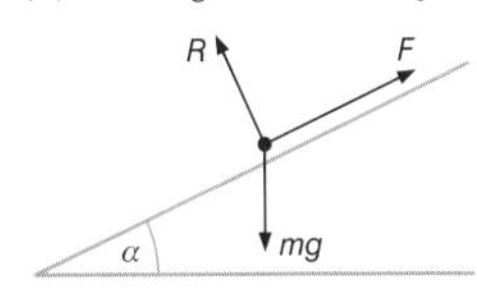

3 If the high-jumper has mass m kg and the moon has mass M kg then the force of attraction on the athlete by the moon is given by $\frac{GmM}{R^2}$.
This is equal to his weight on the moon (say mg_1 where g_1 denotes the acceleration due to gravity on the moon).

$$g_1 = \frac{GM}{R^2} = 1.64$$

So, on the moon, the acceleration due to gravity is about $\frac{1}{6}$ that on the Earth and an athlete would be able to jump about six times as high.
The lunar record would thus be approximately $6 \times 2.3 = 13.8\,\text{m}$!

4 The forces acting on the two masses are shown in the diagram, where F denotes the friction force.
Since mass B is not moving, resolving vertically for it gives:
$T = 5g = 49\,\text{N}$
Since mass A is also not moving, resolving horizontally for it gives:
$F = T = 49\,\text{N}$

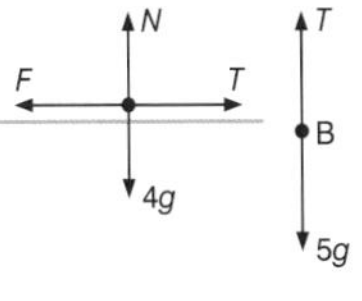

5 Since the system is in equilibrium the forces must be in balance.
Vertical forces: $T_1\sin25^\circ + T_2\sin50^\circ = 3g$
Horizontal forces: $T_1\cos25^\circ = T_2\cos50^\circ$
Solving: $T_1 = 19.56\,\text{N}$, $T_2 = 27.58\,\text{N}$.

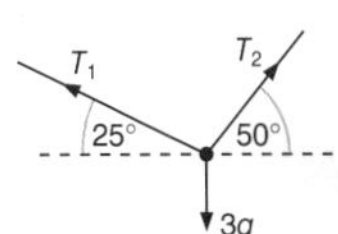

6 Since the box is just about to move the system is in balance.
Perpendicular to slope: $R = 2g\cos30°$
Parallel to slope: $F + 2g\sin30° = 19.6$
Friction law: $F = \mu R$
Substituting from the above gives:
$$\mu = \frac{19.6 - 2g\sin30°}{2g\cos30°} = \frac{19.6 - g}{g\sqrt{3}} = \frac{1}{\sqrt{3}}$$

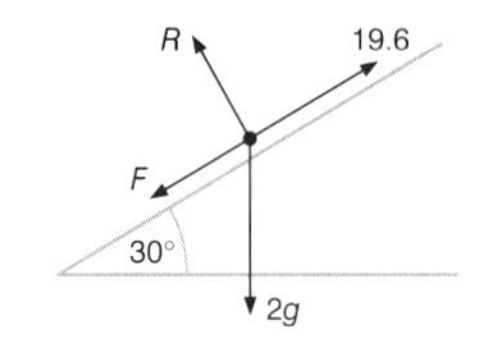

7

Force (N)	**Vector**
2	$2\mathbf{i} + 0\mathbf{j}$
4	$4\cos60°\mathbf{i} + 4\sin60°\mathbf{j}$
1	$0\mathbf{i} - 1\mathbf{j}$
3	$-3\cos30°\mathbf{i} - 3\sin30°\mathbf{j}$
F	$a\mathbf{i} + b\mathbf{j}$

The resultant is $(a + 1.402)\mathbf{i} + (b + 0.964)\mathbf{j}$ so $\mathbf{F} = -1.402\mathbf{i} + 0.964\mathbf{j}$
Magnitude = 1.73 N
Direction = $\tan^{-1}\frac{0.964}{1402} = 34.5°$

8 **(a)**

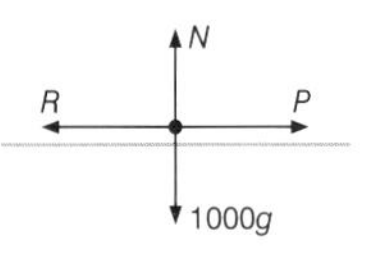

(b) At maximum speed the motion is uniform so the force system balances and $P = R$.

(c)

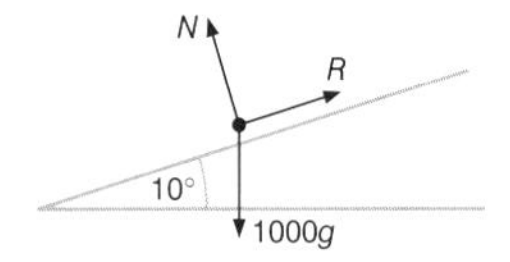

$1000g\sin10° = 1702\text{ N} = R$
At maximum speed $v^2 + 20v - 1702 = 0$
Solving for v gives $v = 32.4$ or -52.4.
The maximum speed reached is 32.4 m s^{-1}.

Check yourself answers

KINEMATICS IN ONE DIMENSION (page 67)

1 The particle accelerates uniformly from $50\,\text{m s}^{-1}$ to $100\,\text{m s}^{-1}$, then decelerates uniformly from $100\,\text{m s}^{-1}$ to zero.

For the first part of the journey, constant acceleration $= \dfrac{100-50}{10-0} = 5\,\text{m s}^{-2}$

For the second part of the journey, constant deceleration $= \dfrac{100-0}{50-10}$
$= 2.5\,\text{m s}^{-2}$.

Initial velocity $= 50\,\text{m s}^{-1}$, maximum velocity $= 100\,\text{m s}^{-1}$, final velocity $= 0\,\text{m s}^{-1}$.
Distance travelled = area of trapezium OABP + area triangle BPQ
$= \frac{1}{2} \times 10 \times (50 + 100) + \frac{1}{2} \times 40 \times 100 = 2750\,\text{m}$

2 **(a)** From the velocity–time graph the gradient is the acceleration $= 0.5\,\text{m s}^{-2}$.
(b) Distance travelled = area under the graph.
(i) 4 m **(ii)** 16 m **(iii)** $0.25t^2$
(c) Distance after 12 seconds = 36 m, in the final 48 seconds he travels $48 \times 6 = 288$ m so the distance travelled in 1 minute is 324 m.

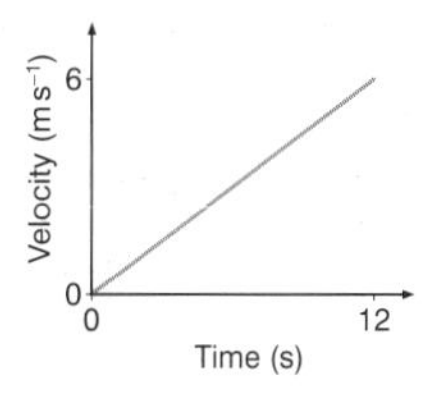

3 $a = g\sin 30° = 4.9$, $u = 0$ and $t = t$ so using $v = u + at$ gives $v = 4.9t\,\text{m s}^{-1}$. The result indicates that the velocity will increase without limit as the time increases. This is unrealistic, experience suggests that there is a limit to the speed attained (consider downhill skiers). Resistance effects have been ignored.

4 For the first part of the journey $a = 0.5$, $u = 0$, $v = 20$, $s_0 = 0$ so using $v^2 = u^2 + 2a(s - s_0)$ gives $20^2 = 0^2 + 2 \times 0.5(s - 0)$
$s = 400\,\text{m}$
To find the time for the first part use $v = u + at$ to give $t = 40\,\text{s}$.
For the second part of the journey $u = 20$, $v = 0$, $s = 120$, $s_0 = 0$ so using $v^2 = u^2 + 2a(s - s_0)$ gives $a = -\frac{5}{3}\,\text{m s}^{-2}$. Now use $v = u + at$ to find $t = 12$.

So the total distance travelled is 520 m, the total time taken is 52 s.

5 Using $v = u + at$ gives $a = 0.35\,\text{m s}^{-2}$ and then $v^2 = u^2 + 2a(s - s_0)$ gives $v = 16.8\,\text{m s}^{-1}$.

Check yourself answers

Motion and Vectors (page 70)

1 Using $\mathbf{r} = \mathbf{r}_0 + \mathbf{u}t + \frac{1}{2}\mathbf{a}t^2$ gives $\mathbf{r} = 0.75t^2\mathbf{i} - t^2\mathbf{j}$.

When $t = 3$, $\mathbf{r} = 6.75\mathbf{i} - 9\mathbf{j}$.
The length of the slide is given by:
$\sqrt{6.75^2 + 9^2} = 11.25\,\text{m}$
The speed of the child at the bottom of the slide is found from $\mathbf{v} = \mathbf{u} + \mathbf{a}t$ giving $\mathbf{v} = 4.5\mathbf{i} - 6\mathbf{j}$ so the speed at the bottom of the slide is $\sqrt{4.5^2 + 6^2} = 7.5\,\text{m s}^{-1}$.

2 Velocity before rebound:
$\mathbf{u} = 6\sin45°\mathbf{i} + 6\cos45°\mathbf{j} = 3\sqrt{2}\mathbf{i} + 3\sqrt{2}\mathbf{j}$
Velocity after rebound:
$\mathbf{v} = -5\sin60°\mathbf{i} + 5\cos60°\mathbf{j} = -\frac{5\sqrt{3}}{2}\mathbf{i} + 2.5\mathbf{j}$
Average acceleration during contact:
$\mathbf{a} = \frac{\text{change in velocity}}{\text{time taken}} = \frac{\mathbf{v} - \mathbf{u}}{0.1} = -85.7\mathbf{i} - 17.4\mathbf{j}\,\text{m s}^{-2}$

3 **(a)** For the first 10 seconds the acceleration of the boat is $\mathbf{a} = 1.15\mathbf{i} + 0.96\mathbf{j}$.
Using $\mathbf{r} = \mathbf{r}_0 + \mathbf{u}t + \frac{1}{2}\mathbf{a}t^2$ gives $\mathbf{r} = 0.575t^2\,\mathbf{i} + 0.48t^2\mathbf{j}$,
so after 10 seconds the position is $57.5\mathbf{i} + 48\mathbf{j}$.
The distance travelled is $\sqrt{57.5^2 + 48^2} = 74.9\,\text{m}$.

(b) The velocity at the end of this 10 seconds is $11.5\mathbf{i} + 9.6\mathbf{j}$ and the acceleration is now zero. Using this fact to find the position at the end of a further 30 seconds gives $\mathbf{r} = 402.5\mathbf{i} + 336\mathbf{j}$ and so the speedboat has now travelled a total distance of 524.3 m.

Check yourself answers

NEWTON'S LAWS (page 73)

1 The resultant force on the skier is:
$(100 - 70g\sin20°)\mathbf{i} + (R - 70g\cos20°)\mathbf{j}$
Perpendicular to the slope, the skier's motion does not change so the $\mathbf{j}$-component of the resultant force is zero i.e. $R = 70g\cos20°$.
Applying Newton's second law of motion parallel to the slope gives:
$70a = 100 - 70g\sin20°$
so $a = -1.92\,\text{m}\,\text{s}^{-2}$
The skier accelerates *down* the slope at $1.92\,\text{m}\,\text{s}^{-2}$.

2

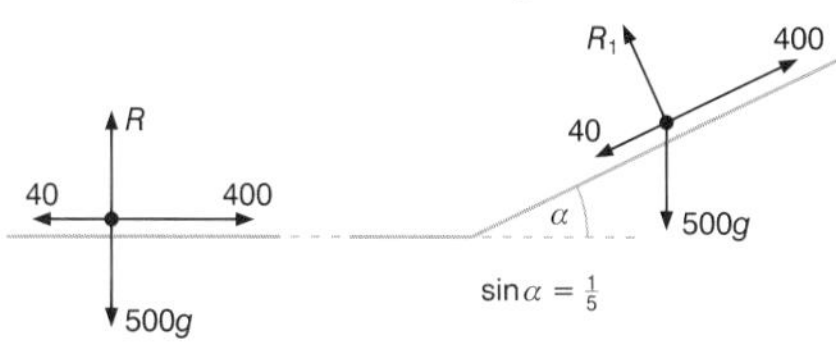

If the acceleration of the car on the horizontal road is a then:
$500a = 400 - 40$, $a = 0.72\,\text{m}\,\text{s}^{-2}$.
Using $v^2 = u^2 + 2as$ to find the velocity at the foot of the slope gives $v = 19.7\,\text{m}\,\text{s}^{-1}$.
If the acceleration of the car up the slope is a_1 then:
$500a_1 = 400 - 500g\sin\alpha - 40$, $a_1 = -1.24\,\text{m}\,\text{s}^{-2}$.
Using $v^2 = u^2 + 2as$ then gives
$s = 156.5\,\text{m}$.

3 The tension is the same on either side of the pulley so both particles have the same acceleration.
Forces on A: $R = 2g\cos30° = 16.97\,\text{N}$
Applying Newton's second law on A:
$T - F - 2g\sin30° = 2a$
Applying Newton's second law on B:
$4g - T = 4a$
From the law of friction:
$F = 6.79$
Solving these equations together gives $a = 3.77\,\text{m}\,\text{s}^{-2}$, $T = 24.1\,\text{N}$.

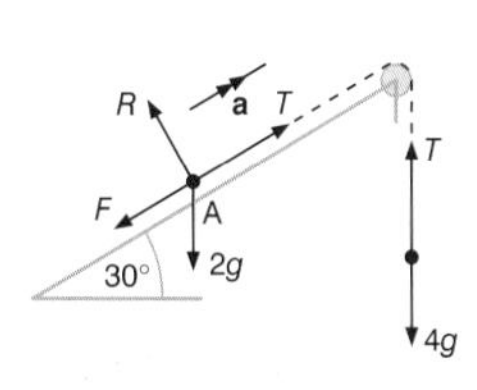

4 Applying Newton's second law on the package:
$T - 5g = 5a$, $T = 54\,\text{N}$.
Model the man and package together and apply Newton's second law.
$R - 65g = 65a$, $R = 702\,\text{N}$.

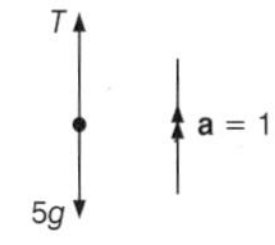

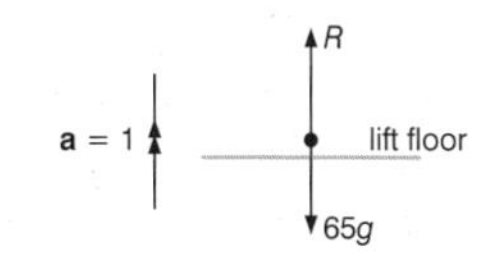

Check yourself answers

Momentum, Collisions, Moments (page 76)

1 I = change in momentum
$= 0.2 \times (-4) - 0.2 \times 6 = -2\,\text{N s}$
The magnitude of the impulse is 2 N s.
Assuming the contact force is constant, $I = Ft$ which gives the magnitude of the contact force as $\frac{2}{0.1} = 20\,\text{N}$.

velocity before $u = 6$
velocity after $v = -4$

2 If V is the initial velocity of the bullet, by conservation of momentum: $0.03V = 8.03 \times 5$
$V = 1338\,\text{m s}^{-1}$

3 **(a)**

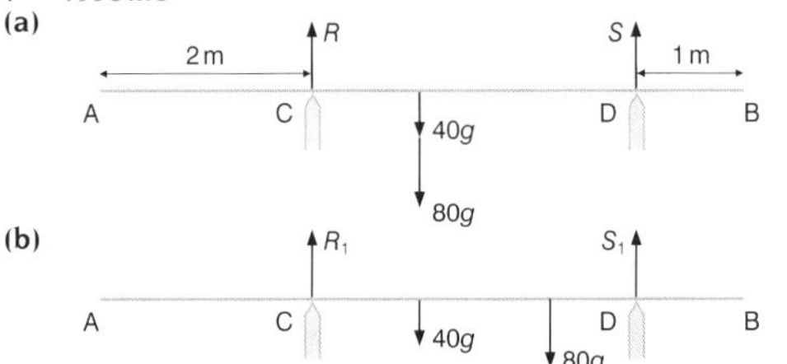

(a) Since the system rests in equilibrium then the total force on the system is zero so $R + S = 120g$, and the total moment on the system is zero.
Taking moments about C gives
$(40g + 80g) \times 1 = S \times 3$, $S = 40g$ and so $R = 80g$.
The reaction at C is $80g$ and the reaction at D is $40g$.

(b) Suppose the man moves a distance x towards B. The system is still in equilibrium:
$R_1 + S_1 = 120g$
Taking moments about C: $40g \times 1 + 80g \times (1 + x) = S_1 \times 3$
$$S_1 = \frac{120g + 80gx}{3}$$
At the instant when the plank is about to turn, contact at C will be lost i.e. $R_1 = 0$.
Thus $S_1 = 120g$ giving $x = 3\,\text{m}$.
The man can just reach the end B before the plank topples.

4 The system is in equilibrium so the total force and the total moment equal zero.
$F = N$ and $R = mg$
Taking moments about the foot of the ladder:
$mgl\cos\theta = N \times 2l\sin\theta$
$F = \mu R \Rightarrow \frac{1}{2}mg\cot\theta = \mu mg \Rightarrow \cot\theta = 2\mu$
Solving for $\tan\theta$ gives $\tan\theta = \dfrac{1}{2\mu}$.

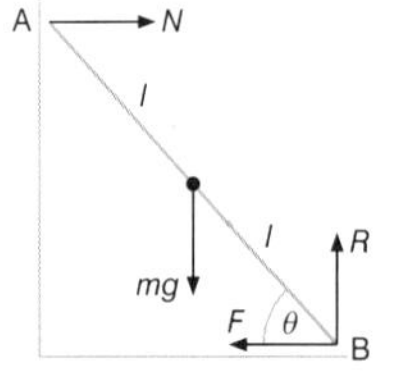

Check yourself answers

Projectiles (page 78)

1 The position of the projectile is give by $x = 30\cos30°t$, $y = 30\sin30°t - 0.5gt^2$ and the corresponding velocities are $u = 30\cos30°$ and $v = 30\sin30° - gt$. After 2 seconds the position is (51.96, 10.4) and the velocity components are $u = 25.98$, $v = -4.6$. The magnitude of the velocity is $26.38\,\text{m s}^{-1}$.
The negative vertical component of the velocity indicates that the particle is moving downwards.

2 The horizontal and vertical velocity components are $u = 350\cos\alpha$ and $v = 350\sin\alpha - 9.8t$. The vertical displacement of the bullet at time t is $y = (350\sin\alpha) - 4.9t^2$. $\sin\alpha = \frac{3}{5}$, $\cos\alpha = \frac{4}{5}$.
At a height of 1250 m, $1250 = 210t - 4.9t^2$ so:
$t = 7.14$ s and $t = 35.71$ s (up and down).
When $t = 7.14$, $u = 280$ and $v = 140$, so the speed is
$\sqrt{280^2 + 140^2} = 313\,\text{m s}^{-1}$.

The angle of direction is $\tan^{-1}\frac{v}{u} = 26.6°$.

The speed is $313\,\text{m s}^{-1}$ and the angle of direction is 26.6°.

3 The vertical displacement is given by $y = (V\sin\alpha)t - 0.5gt^2$ and the vertical velocity component is given by $v = V\sin\alpha - gt$.
The projectile reaches its maximum height when $v = 0$ giving

$$t = \frac{V\sin\alpha}{g}.$$

Substituting in y:

$$y = (V\sin\alpha)\frac{V\sin\alpha}{g} - 0.5g\frac{V^2\sin^2\alpha}{g}$$

$$= \frac{V^2\sin^2\alpha}{2g}$$

which is the required result.

4 The trajectory of the particle is:

$$y = 49 + x\tan45° - \frac{gx^2\sec^2 45°}{2 \times 14^2}$$

$$y = 49 + x - 0.05x^2$$

Solving gives x (horizontal distance) = 42.86 and $x = -22.86$.
The particle lands 42.86 m from the cliff.

5 Relative to the point from which the ball is thrown, its horizontal and vertical displacements are, respectively:

$$x = ut \quad \text{and} \quad y = vt - \tfrac{1}{2}gt^2$$

Hence $60 = ut$ and $0 = vt - \frac{1}{2}gt^2$

Thus $t = \frac{60}{u}$ so $\frac{60}{u} = \frac{2v}{g} \Rightarrow uv = 30g \Rightarrow uv = 294.$

Check yourself answers

Handling Data (pages 85–86)

1 Unit is 1

3	7
4	3 8
5	0 2 6 9
6	0 3 4 5 5 8
7	2 5 6 9 9
8	3 8

The median is the middle value. This is the mean of the 10th and 11th values, which are 64 and 65. Therefore the median is 64.5. The data are slightly negatively skewed.

2 You must use a weighted mean.

$$\bar{x}_w = \frac{1\times 56 + 2\times 62 + 3\times 65 + 4\times 59}{1+2+3+4} = \frac{611}{10} = 61.1\%$$

3 Here $n = 18$, $\Sigma x = 270$, $\Sigma x^2 = 4150$. Therefore $\bar{x} = \frac{\Sigma x}{n} = \frac{270}{18} = 15$

$$\sigma^2 = \frac{\Sigma x^2}{n} - \bar{x}^2 = \frac{4150}{18} - 15^2 = 5.556 \text{ to 3 d.p.}$$

4 Here $\bar{x} = \frac{\Sigma fx}{\Sigma f} = 152.235$ to 3 d.p. and

$$\sigma^2 = \frac{\Sigma fx^2}{\Sigma f} - \bar{x}^2 = \frac{394\,016}{17} - \left(\frac{2588}{17}\right)^2 = 1.827 \text{ to 3 d.p.}$$

5 Adjust the classes to make the data truly continuous and then calculate the frequency density for each class.

Class boundaries	Number	Class width	Frequency density
0–(5.5)	35	5.5	6.36
5.5–(10.5)	42	5	8.4
10.5–(20.5)	29	10	2.9
20.5–(30.5)	45	10	4.5
30.5–(50.5)	26	20	1.3
50.5–75.5	11	25	0.44

The histogram can now be drawn.

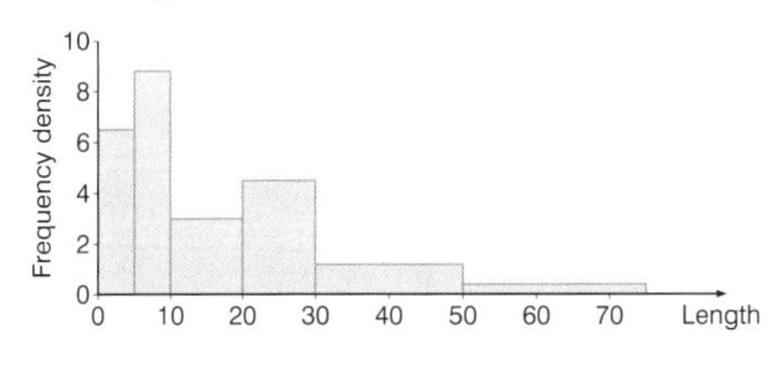

6 Calculate the percentage cumulative frequency, starting from the fact that the total number of students is 50, which means that 50 is equivalent to 100%. Note the addition of the first class.

Upper class boundary	Number of students	Cumulative frequency	Percentage cumulative frequency
59.5	0	0	0
62.5	4	4	8
65.5	9	13	26
68.5	20	33	66
71.5	13	46	92
74.5	4	50	100

The percentage cumulative frequency polygon can be drawn.
The median is the 50% value and is 67.3 from the graph.
The lower quartile is the 25% value and is 65.2 and the upper quartile is the 75% value and is 69.9. The IQR = 4.7.

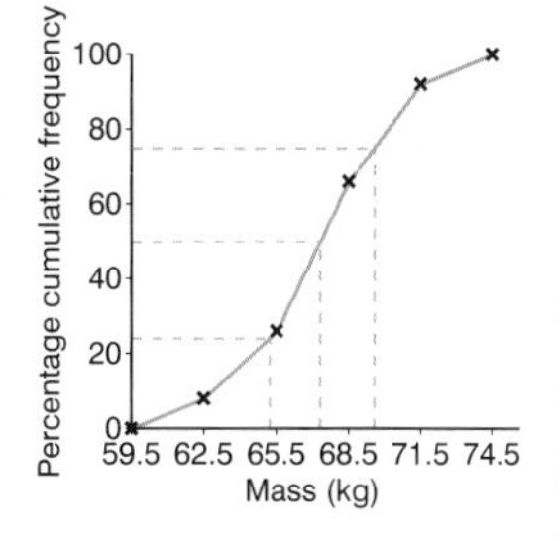

7 The median for this data set is 154 cm, the lower quartile is 135 cm, the upper quartile is 172 cm, the smallest value is 105 cm and the largest value is 196 cm. The box-and-whisker plot is constructed from these figures.

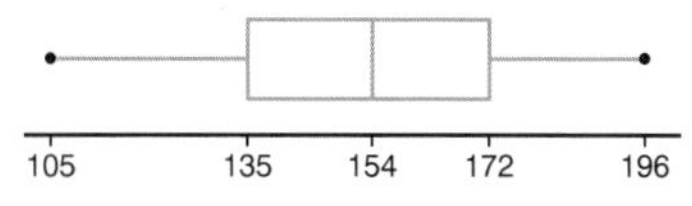

8 A box-and-whisker plot for this data set shows that the data are skewed so therefore the best measure of location is the median and the best measure of spread is the IQR.
median = 8, IQR = 14 – 5 = 9

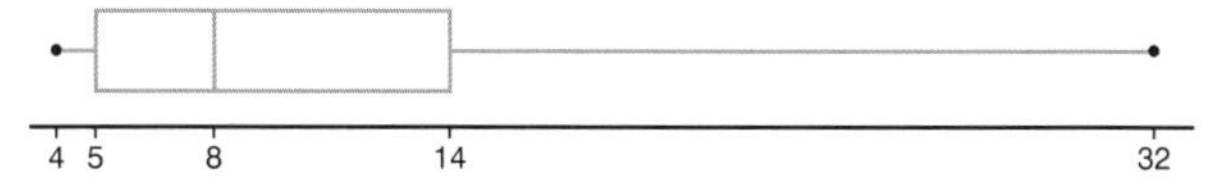

9 $E(X) = \Sigma x \times P(X = x) = 1 \times 0.1 + 2 \times 0.1 + 3 \times 0.3 + 4 \times 0.3 + 5 \times 0.2 = 3.4$

Check yourself answers

PROBABILITY (page 90)

1 As two events are independent, the probability is $\frac{13}{52} \times \frac{1}{6} = \frac{1}{24}$

2 P(at least one tail) = 1 – P(no tails) = $1 - \frac{1}{2} \times \frac{1}{2} = \frac{3}{4}$

P(multiple of 2) = $\frac{2}{5}$

So probability of at least one tail and multiple of 2 $= \frac{3}{4} \times \frac{2}{5} = \frac{3}{10}$

3 For both sweets to be green the probability is $\frac{40}{90} \times \frac{39}{89} = 0.195$

For the sweets to be different colours they could be taken out in the order RG ($\frac{50}{90} \times \frac{40}{89} = 0.2497$) or GR($\frac{40}{90} \times \frac{50}{89} = 0.2497$).

The probability they are different colours is $0.2497 + 0.2497 = 0.499$.

4 **(a)** P(both red) = $\frac{26}{52} \times \frac{26}{32} = \frac{1}{4}$

(b) P(different colours) = P(BR) + P(RB) = $\frac{26}{52} \times \frac{26}{52} + \frac{26}{52} \times \frac{26}{52} = \frac{1}{2}$

(c) P(both are picture cards) = $\frac{12}{52} \times \frac{12}{52} = \frac{9}{169}$

(d) P(first is ace and second is black) = $\frac{4}{52} \times \frac{26}{52} = \frac{1}{26}$

5 **(a)** MUMMY has three Ms so the number of distinct arrangements is $\frac{5!}{3!} = 20$.

(b) LIZZIE has two Zs and two Is so the number of distinct arrangements is $\frac{6!}{2!2!} = 180$.

(c) TIGGER has two Gs so the number of distinct arrangements is $\frac{6!}{2!} = 360$.

6 This is equivalent to selecting seven from 52 when the order does not matter so the number of combinations is

$$^{52}C_7 = \frac{52!}{7!45!} = \frac{52 \times 51 \times 50 \times 49 \times 48 \times 47 \times 46}{1 \times 2 \times 3 \times 4 \times 5 \times 6 \times 7} = 133\,784\,560$$

Check yourself answers

Bivariate Data, Normal Distribution (page 94)

1 **(a)** $\overline{x} = 3$, $\overline{y} = 31.4$, $s_x = 1.414$, $s_y = 6.41$, $s_{xy} = 7.8$

(b) From the scatter plot there is some indication of a positive linear relationship. This is verified by $r = \dfrac{s_{xy}}{s_x s_y} = \dfrac{7.8}{1.414 \times 6.41} = 0.8606$.

(c) The equation of the line of best fit is given by

$\hat{y} - \overline{y} = \dfrac{s_{xy}}{s_x^{\ 2}}(x - \overline{x}) \Rightarrow \hat{y} - 31.4 = \dfrac{7.8}{2}(x - 3) \Rightarrow y = 3.9x + 19.7$

2

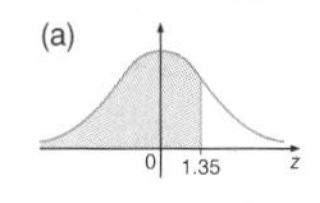

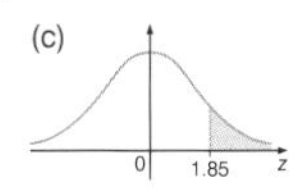

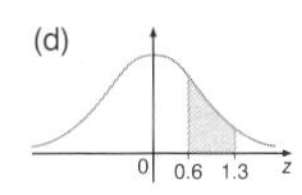

(a) Using a table of the normal distribution $P(z < 1.35) = 0.9115$

(b) Using the symmetry of the normal curve $P(z < -1.0) = 1 - P(z < 1.0)$ $= 1 - 0.8413 = 0.1587$

(c) $P(z > 1.85) = 1 - P(z < 1.85) = 1 - 0.9678 = 0.0322$

(d) $P(0.6 < z < 1.3) = P(z < 1.3) - P(z < 0.6) = 0.9032 - 0.7258 = 0.1774$

3 $\mu = 10.1$, $\sigma = 0.5$, standard deviation $z = \dfrac{x - \mu}{\sigma}$

(a) $x = 10.3 \Rightarrow z = \dfrac{10.3 - 10.1}{0.5} = 0.4$, $P(x < 10.3) = P(z < 0.4) = 0.6554$

(b) $x = 9.8 \Rightarrow z = \dfrac{9.8 - 10.1}{0.5} = -0.6$, $P(x > 9.8) = P(z > -0.6) = 0.7258$

(c) $9.9 < x < 10.4 \Rightarrow -0.4 < z < 0.6$

$P(9.9 < x < 10.4) = P(-0.4 < z < 0.6) = P(z < 0.6) - P(z < -0.4)$
$= 0.7258 - (1 - 0.6554) = 0.3812$

(d) Let the lengths be x_1 and x_2 with corresponding z-values of z_1 and z_2.
$P(z_1 < z < z_2) = 0.95$ and $z_1 = -z_2$ by symmetry and
$P(z < z_2) = 0.5 + 0.475 = 0.975 \Rightarrow z_2 = 1.96$.

Thus $\dfrac{x - 10.1}{0.5} = \pm 1.96$

$\Rightarrow x = 10.1 \pm 0.98 \Rightarrow x = 9.12, 11.08$.

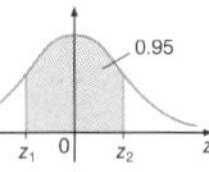

The lengths required are between 9.1 cm and 11.1 cm (1 d.p.)

4 $\mu = 55$, $\sigma = 15$, $z = \dfrac{x - \mu}{\sigma}$

(a) $P(x > 70) = P(z > 1)$
$= 0.159 \approx 16\%$

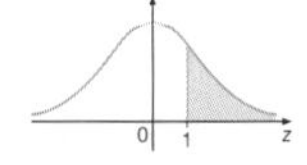

(b) $P(x < 30) = P(z > -1.67)$
$= 0.048$ using symmetry

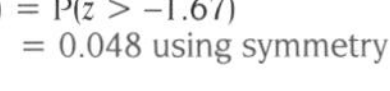

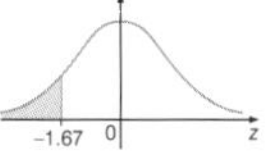

(c) $P(30 < x < 39) = P(-1.67 < z < -1.07) = 0.095$

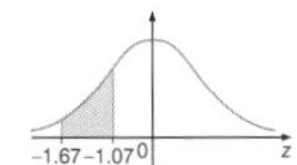

Index

Index